POCKET EINSTEIN

10 Short Lessons in
Space Travel

POCKET EINSTEIN

10 Short Lessons in
Space Travel

Paul Parsons

Michael O'Mara Books Limited

First published in Great Britain in 2020
by Michael O'Mara Books Limited
9 Lion Yard
Tremadoc Road
London SW4 7NQ

A CIP catalogue record for this book is available from the British Library.

Papers used by Michael O'Mara Books Limited are natural, recyclable
products made from wood grown in sustainable forests. The
manufacturing processes conform to the environmental regulations of
the country of origin.

ISBN: 978-1-78929-221-3 in hardback print format
ISBN: 978-1-78929-240-4 in ebook format

1 2 3 4 5 6 7 8 9 10

www.mombooks.com

Designed and typeset by Ed Pickford
Illustrations by David Woodroffe
Front cover illustration by Siaron Hughes

Printed and bound by CPI Group (UK) Ltd, Croydon, CR0 4YY

MIX
Paper from
responsible sources
FSC® C020471

CONTENTS

INTRODUCTION

Among my earliest childhood memories are some vivid images of a magical world. Not Narnia or Hundred Acre Wood, but a red-brown planet called Mars. It was 1976, the year Concorde took to the skies, twenty-nine nations boycotted the Olympic Games in protest at South African apartheid, and VHS video cassettes first went on sale. It was also the year that, on 20 July – just over a month before my fifth birthday – NASA's Viking 1 lander touched down on Mars and returned the first ever pictures from its surface. I remember watching the news reports showing the spacecraft's landing foot planted firmly in the rock-strewn Martian soil. And then the amazing vistas of a rusty-red world with a pale-salmon sky. I was enthralled.

The Viking imagery made space feel real. More tangible. There were already pictures taken by astronauts in orbit, or on the moon, or images of the distant planets like Jupiter captured by astronomers through their telescopes. And these were incredible. But they were also alien, and abstract – so much so that it was difficult to relate what you were seeing to the real world. The images returned from Viking 1, and its sister craft Viking 2, which landed on Mars two months later, showed us a world that could, at a glance, be mistaken for Earth. You could imagine standing there; you could almost reach out and touch it.

Today, we're standing at a pivotal juncture in the human exploration of space. For the first time in almost fifty years, human astronauts are poised to return to the moon and, it's hoped, will continue on to Mars and even further afield. Humans last set foot on the moon in 1972, and haven't been beyond low Earth orbit since. The new flurry of activity we're about to see stems from renewed interest in the exploration and scientific study of the moon by American space agency NASA – which even has plans to put a space station in lunar orbit – as well as the rapid expansion of the private space launch and space tourism industries.

And yet it's not just about flags and footprints. The next decade will see a wave of new robotic spacecraft take flight to explore the planets and moons of our solar system. New lander missions to Mars will attempt to sniff out the chemical and geological signatures of past life on the planet, and will deploy a helicopter scout to survey the Red Planet from the air. The formidable James Webb Space Telescope – the successor to the Hubble Space Telescope – will begin its primary mission, studying the birth of planets, stars, and the first generation of galaxies (groups of many stars, like our own Milky Way) that condensed billions of years ago from the hot soup of the early universe. And there will be new missions heading to the outer realms of the solar system, studying the gas-giant worlds Jupiter and Saturn and their moons, some of which may have oceans of liquid water lurking beneath their frozen surfaces. The findings of these missions may, like Viking before them, inspire the next generation of astronomers and space scientists.

This book is a brief tour around the science of space travel, from its beginnings in the nineteenth century with the musings of Russian rocket pioneer Konstantin Tsiolkovsky, through the East–West Space Race and the pomp and glory of the moon landings, to the Space Shuttle and the International Space Station, and the amazing future concepts that could see human explorers travelling to other star systems or colonizing other planets.

Exploring space is part and parcel of understanding the universe and our place in it, and it may well prove vital to the ultimate survival of our species. So climb aboard the capsule, strap yourself in, and get comfortable. We're about to make the next giant leap.

Paul Parsons

01 HOW FAR WE'VE COME

'To be the first to enter the cosmos, to engage, single-handed, in an unprecedented duel with nature – could one dream of anything more?'

YURI GAGARIN (1961)

The feeling of anxiety in the pit of your stomach has grown close to nausea, and your mouth is parched. But who can blame you? You're strapped into a capsule atop a space rocket taller than an office block. And now it's ready to launch, laden with enough fuel to detonate with the force of a small nuclear weapon.

You've been in your seat for nearly three hours, waiting patiently as mission controllers run test after test on the spacecraft's systems. All the while, you've rehearsed the task ahead of you. Running through different contingencies in your mind. Praying a little. Now the countdown is into its final moments.

Before you can ruminate further, there's a jolt followed by a heavy rumble. The cabin lurches from one side to the other and then begins to shake violently as the main engines ignite. As they throttle up to maximum power, the G-force presses down like a great weight on top of you, forcing you back into your seat as millions of pounds of thrust hurl you and the rocket up towards the sky.

A little over a minute into the flight and you're already travelling faster than the speed of sound. As you continue to accelerate, the howl of the wind outside becomes eerily audible as it rushes over the capsule's thin outer skin. A short time later, there's a loud bang and you feel a hard kick in the back as the rocket's first stage is jettisoned and the second-stage engines light up. You're now 60 km (37 miles) up, moving at nearly 10,000 km per hour (6,214 mph), and still accelerating. Gradually, the sky fades to black and, as the atmosphere grows thinner, the wind and the pounding vibrations die away.

You've now reached orbit, flying 300 km (186 miles) above the surface of the Earth at a speed of approximately 28,000 km/h (17,400 mph) – that's 7.8 km (4.8 miles) every single second, more than twenty times the speed of sound.

Abruptly, the engine shuts off and you get your first taste of zero gravity, floating out of your seat until the straps of your harness snap taught to keep you in place. Your stomach isn't so fortunate. It feels like it's in your throat, as if you're permanently driving over the crest of a humpback bridge. Glancing out of the window, you see the Earth wheeling below – a cloud-swathed pearly-blue sphere set against the inky darkness beyond. Congratulations: you have just become the latest human explorer to venture into the beguiling realm of outer space.

Stranger than fiction

Fiction is a handy device for conveying the incredible, but it's also how humanity's long trek into space actually began.

Perhaps the first tale of a journey beyond the confines of Earth was written by the Assyrian writer Lucian of Samosata in the year 160 CE. Called *A True Story*, it tells of a group of sailors whose ship is blown off course by a storm and caught up in a terrible whirlwind that scoops up the vessel and deposits it on the moon. Here, the crew discover weird alien lifeforms, some loyal to the moon and others to the sun, locked in a savage conflict. Though, ultimately, peace prevails.

The German astronomer Johannes Kepler is best known for formulating the laws of planetary motion that govern the behaviour of our solar system. However, in the early seventeenth century he penned a science-fiction

> Don't tell me that man doesn't belong out there. Man belongs wherever he wants to go — and he'll do plenty well when he gets there.
>
> WERNHER VON BRAUN
> (1958)

story called *Somnium* (*A Dream*). The manuscript described a trip to the moon, and what the heavens might look like from this very different vantage point according to his sun-centred view of the solar system (revolutionary at the time).

Both these stories invoked magic to transport the intrepid explorers into space. It wasn't until the late nineteenth century that fiction started to bridge the gap between fanciful notions of space travel and actual workable technologies. In Jules Verne's 1865 classic *From the Earth to the Moon*, and its sequel *Around the Moon*, a crew of three astronauts are shot into space by a giant cannon. Oddly enough, the cannon was situated in Florida, which would later become the site of Cape Canaveral. Later again, H. G. Wells wrote about astronauts propelled to the moon by an anti-gravitating substance called Cavorite, in his novel *The First Men in the Moon*.

> **All civilizations become either spacefaring or extinct.**
>
> CARL SAGAN (1994)

It was around this time that the early steps were taken towards turning space travel into science fact. In 1903, Russian scientist Konstantin Tsiolkovsky published an article titled 'The Exploration of Cosmic Space by Means of Reaction Devices', in which he detailed how rockets might be used to leave Earth and fly between the planets.

On 17 December that same year, in Kitty Hawk, North Carolina, brothers Wilbur and Orville Wright made the first flight of a heavier-than-air powered aircraft. The Wright Flyer was a biplane with a wingspan of 12 metres (39 feet), powered by a 12-horsepower petrol engine driving two propellers. It made four flights that day, the longest of which lasted for 59 seconds and covered a distance of 260 metres (850 feet). Hardly long-haul, but it was a start. Humans were no longer confined to their planet's surface.

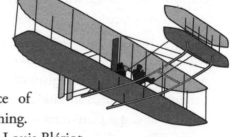

From here, the pace of advancement was astonishing. In 1908, French aviator Louis Blériot flew across the English Channel. And by the outbreak of the First World War, a little over a decade after the Wright Brothers' first flight, aircraft had developed sufficiently to be repurposed as formidable weapons.

Rocketry was progressing, too. Tsiolkovsky's ideas were being applied by the US engineer and inventor Robert Goddard. In 1926, he launched the world's first liquid-propelled rocket from a site in Auburn, Massachusetts. The rocket, named *Nell* and powered by an engine running on petrol and liquid oxygen, rose 12.5 metres (41 feet) into the air in a flight lasting 2.5 seconds. In the 1930s, Goddard developed guided rockets that could break the sound barrier and reach altitudes of almost 3 km (2 miles).

But Goddard wasn't the only one to have taken an interest in Tsiolkovsky's work. In Europe, an enthusiastic young scientist named Wernher von Braun had been appointed technical director of the German Army Rocket Research Group. By 1934, their A2 rocket had almost equalled Goddard's achievements. But von Braun and his team were working on another project, their most ambitious yet. In 1942, they tested the new rocket, called the A4. Measuring 14 metres (46 feet) in height, ten times taller than the A2, its first flight reached an altitude of 189 km (117 miles), making it the first human-made object to reach space.

> **Houston, Tranquility Base here. The Eagle has landed.**
>
> NEIL ARMSTRONG (1969)

The A4 was later fitted with an explosive warhead and renamed the V2 ('V' for Vergeltungswaffe, or Vengeance Weapon). With a range of over 300 km (185 miles), it was the world's first long-range ballistic missile, and it rained down devastation upon Allied cities towards the end of the Second World War.

The Cold War

When the war ended, Russia and America emerged as the dominant powers in the world. And so began the Cold War – a period of attrition between the United States (and its allies) and the Soviet Union, which lasted until the early 1990s.

Essential for dominance in this period was mastery of the new technologies of the time – nuclear energy, which had been demonstrated by the allies to devastating effect at Hiroshima and Nagasaki, and spaceflight, the potential of which the V2 rockets had proven beyond all doubt.

The rivalry between the US and the USSR to conquer space became known as the Space Race. Both sides had captured German V2 rocket hardware, and secured the services of the project's scientists as the war drew to a close.

Under the US's Operation Paperclip, Nazi scientists were offered American citizenship – and a waiver of their war crimes charges. Peenemünde and Mittelbau-Dora, where the V2s were developed and built, were concentration camps employing slave labour in appalling conditions. Over ten thousand people died here, more than were killed by the weapons they built.

Wernher von Braun was among the scientists recruited through Operation Paperclip. His genius was quickly recognized and he became the director of the American rocketry project, first with the army and later at NASA (the National Aeronautics and Space Administration).

Russia was rather less equitable in its dealings, press-ganging Nazi scientists – quite literally at gunpoint – and taking them to work on its own rocket development programme, which was masterminded by the Soviet engineer Sergei Korolev.

Both teams spent the early years developing intercontinental ballistic missiles (ICBMs) – rockets like the V2, but powerful enough to lift bulky nuclear weapons and sling them from one side of the planet to the other. The Soviets built the R-7, which evolved into the Sputnik and Vostok spacecraft, and later the Soyuz, which is still in service today.

Von Braun's team began working at the US Army's White Sands Proving Ground, in New Mexico. They created the first ever two-stage rockets by joining a small WAC Corporal research rocket to the top of a V2. In 1950, one of the later flights in this so-called Bumper-WAC programme became the first rocket to be launched from the newly established Cape Canaveral in Florida.

It wasn't until the late 1950s that the emphasis shifted away from military applications and the Space Race proper got underway. By 1957, both nations announced their intention to send an artificial satellite into orbit around the Earth. The USSR got there first, launching Sputnik 1 on 4 October that year, followed on 3 November by Sputnik 2 carrying the dog Laika, the first animal in orbit.

In December 1957, things went from bad to worse for the US when its first attempt at a satellite launch, Vanguard-TV3, blew up spectacularly on the pad in front of the world's media. They got their act together soon enough though, and on 31 January 1958, the Explorer 1 mission took off flawlessly from Cape Canaveral and reached orbit shortly afterwards, where it remained until 1970.

NASA was established later in 1958, and the following year instigated its Mercury programme to put a human in space atop a Redstone rocket – one of the ballistic missiles developed by von Braun and his team, now based in Huntsville, Alabama.

> **When I first looked back at the Earth, standing on the moon, I cried.**
>
> ALAN SHEPARD (1988)

But the Soviets got there first again. On 12 April 1961, Yuri Gagarin piloted his *Vostok 1* capsule to perform a single orbit of the Earth and returned to Earth shortly afterwards. A further five Vostok flights would follow.

Again, the US scrambled to catch up. On 5 May 1961, Alan Shepard became the first American in space when his Mercury-Redstone spacecraft propelled him on a fifteen-minute flight, reaching an altitude of 187 km (117 miles).

Shepard's flight was 'sub-orbital' – a short hop up into space and back down again. NASA engineers next replaced the Redstone with the more powerful Atlas booster, and on 20 February 1962, John Glenn became the first US astronaut to orbit the Earth, making three complete circuits before his Mercury capsule splashed down in the Atlantic Ocean 4 hours and 55 minutes later.

YURI GAGARIN (1934–68)

Yuri Alekseyevich Gagarin was born in Klushino, a village in Smolensk Oblast, Russia. In the early 1950s, he began training as a foundry worker, while at weekends he learnt to fly. He entered full-time pilot school in 1955 and became a flying officer in the Soviet Air Forces in November 1957. Having shown interest in Russia's uncrewed Luna moon missions, in 1959 he was selected to join the Soviet human spaceflight project. And on 12 April 1961, he became the first human being in space when he piloted the *Vostok 1* mission, making a single orbit of the Earth and returning safely 108 minutes later. *Vostok-1* was Gagarin's only venture into space. Following the flight, he entered Soviet politics, only returning to the space programme to work on spacecraft designs and cosmonaut training. He was killed in an accident on 27 March 1968, when the MiG-15UTI jet fighter he was co-piloting crashed near Moscow.

The first woman to fly in space was Soviet cosmonaut Valentina Tereshkova who, in June 1963, made 48 orbits of the Earth in her Vostok 6 capsule. It wasn't until much later that an American woman went into space, when Sally

Ride flew aboard the Space Shuttle Challenger in June 1983, though women were playing a pivotal role in the US space programme on the ground. Katherine Johnson joined in 1953. One of a number of female mathematicians, she calculated trajectories – including those of Shepard's and Glenn's early flights, and for the Apollo moon project, while software engineer Margaret Hamilton led the MIT programming team that developed all of Apollo's flight software.

We choose to go to the moon

In September 1962, following President John F. Kennedy's declaration to send humans to the moon, NASA launched its Gemini project, to develop the complex skills and techniques needed to pilot a spacecraft to Earth's only natural satellite.

Russian scientists had a head start here as well, having landed the uncrewed Luna 2 mission on the moon's surface back in 1959 – the first artificial object from Earth to set down on another world. The first US robotic lander mission to reach the moon was Ranger 4, which crash-landed in April 1962.

The US project to put humans on the moon was

> The exploration of space will go ahead, whether we join in it or not, and it is one of the great adventures of all time.
>
> JOHN F. KENNEDY (1962)

called Apollo, and it fell to von Braun's team to develop a rocket powerful enough to send a lander with a human crew on the 380,000 km (236,000 mile) trip. The result was the Saturn V, a rocket that when fully fuelled weighed almost 3,000 metric tonnes (3,300 US tons), and stood 111 metres high (364 feet) – 18 metres (59 feet) taller than the Statue of Liberty.

As America soldiered on with Apollo, Russian engineers were working on their own answer to the Saturn V rocket: the N1. However, the project was beset with problems. All four attempted launches of the rocket failed, the second exploding in a fireball so destructive it obliterated the pad and surrounding launch complex. In their haste to beat America to the moon, engineers had forgone fundamental testing of the N1, which might have exposed some of its technical issues earlier.

As it was, America got to the moon first. After a number of crewed and uncrewed test flights, NASA decided it was ready to attempt a human landing in the summer of 1969. On 20 July, the *Apollo 11 Lunar Module Eagle*, carrying astronauts Neil Armstrong and Buzz Aldrin, set down safely in the Sea of Tranquillity – a giant basin on the surface of the moon – and Armstrong became the first human to walk upon the moon's surface.

Another six Apollo missions followed, in which astronauts explored the moon and carried out geological surveys and scientific experiments. The final flight, *Apollo 17*, took place in December 1972. A total of twelve astronauts got to walk on the lunar surface, and neither the US nor any other nation has been back since.

> **Astronauts are inherently insane. And really noble.**
>
> ANDY WEIR
> *The Martian* (2014)

Boldly gone

With the end of the East–West Space Race, the pace of development in crewed space exploration slowed. Apollo had cost an estimated $153 billion (in 2018 US dollars) and now that America had proved its superiority in space the public and political appetite to continue such spending largely evaporated.

Many lament the lack of progress since the 1960s.

And yet, where humans have declined to boldly go, robots have travelled in their place. There's been a surge in the development of robotic spacecraft – which offer a more robust and cost-efficient way to explore outer space. Uncrewed probes have now visited every world in

THE TEN GREATEST SPACE MISSIONS

Name	Launch date	Description
Sputnik 1	4 Oct 1957	The first artificial satellite, blasted into space from Russia's Baikonur Cosmodrome
Vostok 1	12 Apr 1961	First flight of a human being – Russia's Yuri Gagarin – into space
Apollo 11	16 Jul 1969	First crewed landing on the moon, by Neil Armstrong and co-pilot Buzz Aldrin
Voyager	Aug 1977	Two probes gave the first close-up views of Jupiter, Saturn, Neptune and Uranus
COBE	18 Nov 1989	The Cosmic Background Explorer (COBE) returned the first images of the microwave echo from the Big Bang, revolutionising cosmology
Hubble Space Telescope	24 Apr 1990	A powerful space telescope free from the obscuring murk of the Earth's atmosphere
Cassini-Huygens	15 Oct 1997	Revolutionized our understanding of the planet Saturn and its moons
International Space Station	20 Nov 1998	A permanently crewed multi-national platform in Earth orbit
New Horizons	19 Jan 2006	Returned the first close-up imagery of the dwarf planet Pluto
Curiosity	26 Nov 2011	NASA's most successful rover mission to explore the surface of Mars

our solar system, landing on many and returning stunning pictures and scientific data. Meanwhile, space telescopes, flying above the obscuring haze of Earth's atmosphere, have peered further afield to return breathtaking views of the wider universe, and unlock some of its secrets.

In 1973, the remaining Apollo hardware was used to launch the Skylab space station, and for ferrying crews to and from the orbiting platform. Two years later, an Apollo module and a Russian Soyuz spacecraft performed a historic docking in orbit, signifying a new period of détente between the two former rivals.

Russia itself launched a series of space stations, beginning with the Salyut programme in the 1970s, and in 1986, Mir, aboard which several long-duration spaceflight records were achieved that still stand today. Space stations were seen as a testbed through which to investigate the long-term effects of space travel on the human body, and as a springboard from which to explore the wider solar system.

To this end, the US Space Shuttle programme, which started flying in 1981, was conceived as a transportation solution to ferry cargo and people to and from Earth orbit. One of its principal purposes was to facilitate the construction of a new space station, a collaborative effort between nations. All spacecraft thus far had been expendable, only good for a single launch, but the shuttle could be used multiple times, and rather than splashing down in the sea it landed on a runway like an aircraft.

The first components of the International Space Station (ISS) were launched into orbit in 1998. Construction of the station has required more than forty assembly flights by the shuttle, as well as by Russian Soyuz and Proton launch vehicles. Measuring 109 metres (358 feet) on its longest axis (almost the size of a football pitch), the finished space station weighs 420 tonnes (463 US tons) and provides astronauts with an internal living space of 915 cubic metres (32,330 cubic feet).

After the completion of the ISS, the Space Shuttle was retired in 2011. This has meant that for almost a decade now the US has lacked the capability to launch humans into space, instead having to hitch lifts on Russian Soyuz rockets. That's soon to change. NASA is developing a new heavy-lift rocket called the Space Launch System (snappy name to follow), which has its maiden flight due in 2021. This mission, called Artemis 1, will orbit the moon, testing systems and concepts for a planned return to the lunar surface in 2024.

> **Space exploration is a force of nature unto itself that no other force in society can rival.**
>
> NEIL DEGRASSE TYSON
> (2012)

Commercial interests are also getting in on the game. US firm SpaceX has been contracted to operate robotic resupply missions to the ISS since 2012. Now SpaceX, as well as Blue Origin and others, are working on concepts for crewed vehicles with seats available to the highest bidder. They plan to send humans to the moon, as well as on to Mars and beyond.

Space travel not only satisfies our continuing urge to explore, but also aids the advancement of science and may ultimately prove essential for the survival of our species. Today, we stand on the brink of an exciting new era in the human exploration of outer space.

02 HOW TO LEAVE THE PLANET

'Earth is the cradle of humanity, but one cannot live in a cradle for ever.'

KONSTANTIN TSIOLKOVSKY (1911)

In every sense, outer space seems an awfully long way away – a bewildering realm far removed from our own humdrum experience down here on Earth. And yet, it's actually surprisingly close. The boundary marking the edge of space, known as the Kármán line, after the Hungarian-born scientist Theodore von Kármán, who was the first to propose it, is generally acknowledged to be just 100 km (62 miles) up. That's about the same distance, as the crow flies, from Denver to Colorado Springs – roughly an hour's drive.

Obviously, you won't be driving into space any time soon. And you'd have a hard time getting anywhere close even in a modern aircraft. The problem is a force of nature called gravity, which causes all objects in the universe that have mass to be attracted to one another. In many ways it's quite handy – keeping us all stuck to the planet's surface and able to go about our lives without floating off into the void. But it does also make getting into space considerably harder than a road trip. That's because lifting heavy objects against

> **The rockets that have made spaceflight possible are an advance that, more than any other technological victory of the twentieth century, was grounded in science fiction.**
>
> ISAAC ASIMOV (1976)

the pull of gravity is extremely hard work. Hauling a 1-tonne (1.1 US ton) car, or a 30-tonne (33 US ton) spacecraft, up to an altitude of 100 km (62 miles) takes an extraordinary amount of energy.

Jet planes are among the best flying machines we have, but even they're not up to the task of reaching space. They cruise at a maximum altitude of around 10 km (6 miles), and are able to defy gravity using 'lift', an upward force created as air flows over their specially shaped wings. Go much higher than this, and the air becomes so thin that the lift force is no longer enough to hold the plane up.

But there's another problem. Burning fuel involves a chemical reaction called combustion, which combines fuel with oxygen to release energy. If you put a glass over a burning candle, before long the candle goes out, having exhausted all the available oxygen. And it's the same with a jet engine on an aircraft. As the air thins at high altitude there's less and less oxygen around to combust. It doesn't matter how much fuel you carry – without oxygen, it cannot burn.

The solution is simply to bring your own oxygen with you, and that's what rockets, from the Italian word 'rocchetta', meaning bobbin, do. They're capable of generating huge amounts of thrust that can propel them not just beyond the Earth's atmosphere, but across the solar system and even further afield.

Fire arrows

Rockets were invented in China, in or around the eleventh century. Essentially crude fireworks, they were powered by gunpowder, which had been accidentally discovered by Chinese alchemists. It's made by mixing together charcoal, sulphur and saltpetre (aka potassium nitrate). Saltpetre is a powerful oxidizing agent, which gives off oxygen when heated, causing the charcoal and sulphur to burn at a furious rate. The early Chinese rockets, called 'fire arrows', were used to devastating effect in battle by the armies of the Song dynasty, and they were later wielded by the Mongols in their conquests during the thirteenth century.

In the late eighteenth century, the technology was adopted by the Mysorean army in India, which used rockets with fearsome blades attached to defend against British forces during the aggressive exploitation of the region by the British East India Company.

Inevitably, some of these weapons were captured by the British. And these became the inspiration for a programme of military rocket research at the Royal Arsenal, in Woolwich, led by the English inventor and politician William Congreve. Reverse-engineering the Indian technology, he built a range of rockets, consisting of iron cylinders stabilized with long sticks – much like a modern firework (albeit a lot bigger). Indeed, Congreve's largest rockets weighed hundreds of kilograms, measured around 9 metres (30 ft) in length and had a range of more than 3 km (1.8 miles). Each carried an explosive or incendiary warhead, although these often went off in flight.

Congreve successfully demonstrated his first rockets in September 1805. However, they were quite inaccurate. British engineer William Hale went some way towards remedying this later in the nineteenth century. He removed the rockets' sticks and instead angled the direction of the engine exhaust nozzles, so as to make the rocket spin as it flew – in much the same way that a rifle bullet spins to prevent it veering off course.

KONSTANTIN TSIOLKOVSKY (1857–1935)

Konstantin Eduardovich Tsiolkovsky was born in Ryazan Oblast in the western Russian Empire on 17 September 1857. Due to a hearing problem brought on by scarlet fever, he was mainly home schooled. This allowed him scope to pursue the topics that interested him most, and as a teenager he began to contemplate the feasibility of space travel.

He later earned an income from teaching, and spent his spare time conducting his own scientific research, authoring many scientific papers. From the 1890s, he developed the theory of rocketry, applying the established laws of physics to determine the behaviour of a rocket in flight. He wrote up his findings in the 1903 work *Exploration of Outer Space by Means of Rocket Devices,* which served as an inspiration to the engineers who later made rocket travel a reality. Largely a recluse, Tsiolkovsky spent most of his adult life living in a log house near the Russian city of Kaluga, 150 km (93 miles) south-west of Moscow. He died in Kaluga on 19 September 1935, following surgery for stomach cancer.

The potential of these deadly instruments of war for spaceflight didn't become clear until the early twentieth century, when explored by the Russian genius Konstantin Tsiolkovsky.

Action and reaction

Tsiolkovsky turned rocketry into a science, governed by rigorous mathematics – allowing the behaviour of a rocket to be accurately predicted and studied. It was Tsiolkovsky's research that transformed rockets from crude weapons into the vehicles that would eventually take human beings into space.

Perhaps his greatest contribution was the Tsiolkovsky rocket equation, which he published in 1903. It tells you the increase in speed that a rocket can achieve in terms of its initial (fully fuelled) mass, its empty mass (with all fuel burned) and the speed of the exhaust jet that's thrown out by the engine. This total boost in speed created by burning all the rocket's fuel is known as 'delta-v' – 'v' for velocity and 'delta', the mathematical shorthand for 'a change in'.

Tsiolkovsky arrived at his equation from the work of Sir Isaac Newton, the great British physicist. Newton had formulated three succinct laws describing the behaviour of a moving object. The first said that the object will remain at rest or in a state of uniform motion unless acted on by a force. The second said that if the object has mass m and is acted on by a force F, then it will pick up speed at a rate a, such that $F = ma$. But it was

> I just have my own attitude. I'm out here to get the job done, and I knew I had the ability to do it, and that's where my focus was.
>
> ANNIE EASLEY, NASA scientist (2001)

Newton's third law that Tsiolkovsky was most interested in. This says that for every action there is an equal and opposite reaction. Sit on a chair and the chair pushes back to prevent you crashing to the floor.

It's the reason why a rifle kicks back against your shoulder when you fire it – one force pushes the bullet forward, while an equal and opposite force pushes the body of the gun back towards you. Newton's second law then explains why, despite being acted on by the same force, the tiny bullet shoots away at great speed, while the much heavier gun moves back towards you relatively slowly. Similarly, the exhaust from a rocket is low mass but emerges from the engine at very high speed, causing the relatively large mass of the rocket to gradually gather speed in the opposite direction.

Stack 'em up

Tsiolkovsky's equation says that the maximum speed achievable by a rocket increases with the speed of its exhaust. For example, a rocket with 90 per cent of its launch-pad mass taken up by its fuel load can muster a delta-v equal to 2.3 times the speed of the exhaust gases. That means that if the exhaust is moving at a speed of 2,500 metres per second (9,000 kph, or 5,590 mph – achieved by some rockets today) then the final speed of the rocket is 5,750 metres per second (m/s) – equivalent to 20,700 kph (12,860 mph).

That might seem impressive, and it will likely get you above the Kármán line. However, it's not sufficient to reach orbit around the Earth – let alone get you further afield. According to Isaac Newton's law of gravity, orbiting the Earth requires getting up to an altitude of at least several hundred kilometres above the planet's surface and, even after allowing for gravity pulling you back, still be moving at 7,800 m/s (28,080 kph, or 17,450 mph).

To get round this, Tsiolkovsky asked, what if the payload of the rocket – the 10 per cent of the launch mass that isn't fuel – is itself another, albeit smaller, rocket? When the first rocket has finished burning, it is jettisoned and the smaller rocket, now already travelling at 5,750 m/s (20,700 kph, or 12,860 mph) ignites its own engines. When these have also exhausted their fuel, the payload has gained an additional 5,750 m/s, taking its final speed to 11,500 m/s (41,400 kph, or 25,875 mph) and adding more stages increases the final speed further. The idea became known as 'multi-staging'. Tsiolkovsky showed that a multi-stage rocket can always ferry a higher overall payload into space than a single-stage rocket of the same launch mass.

> **If there is a small rocket on top of a big one, and if the big one is jettisoned and the small one is ignited, then their speeds are added.**
>
> HERMANN OBERTH (1967)

The drawback is that the amount of useful payload that can be carried is reduced. A rocket made up of *n* stages, each of which individually can lift 0.1 of its total mass as payload, can lift a payload equal to 0.1^n of the rocket's overall mass. For a two-stage rocket that's 0.01, for a three-stage rocket it's 0.001, and so on. That means that a three-stage rocket with a total fuelled mass on the launch pad of 50,000 kg (110,230 lb) – equal to 50 metric tonnes (55 US tons) – can carry an actual useful payload of just 50 kg (110 lb).

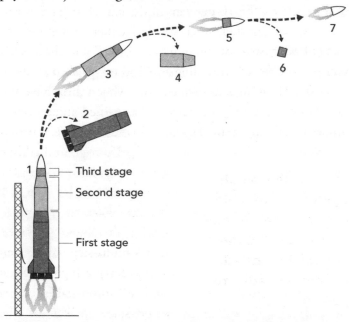

And that's why launching substantial payloads into space requires truly immense multi-stage rockets. Multi-staging was the strategy adopted by NASA in the Apollo moon programme of the 1960s, with the three-stage Saturn V, which remains the largest rocket ever flown. It truly was a behemoth, over 110 metres tall and weighing 3,000 tonnes (3,307 US tons) – 2,870 tonnes (3,164 US tons) of which was fuel and oxidizer. The Saturn V stacked its stages one on top of the other (known as 'serial' staging), but some more recent rockets fire their stages simultaneously (called 'parallel' staging), the two large side boosters on the Space Shuttle being a well-known example.

Rocket science

Modern rocket engines come in several principal types. The most basic, like their early forebears, are powered by solid fuels, similar to gunpowder. The boosters strapped to either side of the Space Shuttle are solid rockets, running on aluminium powder that is burned using the oxidizer ammonium perchlorate. They're potentially hazardous – like Roman candles, once lit, they cannot be switched off, or even throttled down. On the other hand, liquid-fuelled engines, of the sort developed by Robert Goddard (see previous chapter) are more controllable. The price paid for this, however, is additional complexity (and thus increased potential for failure) in the form of pumps, and fuel plumbing, as well as injectors that ensure the fuel gets thoroughly mixed before burning.

Liquid rockets usually store their fuel and oxidizer separately – they are what's known as 'bi-propellant' engines. For example, the first stage of the Saturn V rocket

> **What is it that makes a man willing to sit up on top of an enormous Roman candle ... and wait for someone to light the fuse?**
>
> TOM WOLFE (1979)
> from his novel *The Right Stuff*

that carried the Apollo astronauts to the moon had two fuel tanks, one for kerosene (essentially, jet fuel) and the other filled with pure liquid oxygen to burn it. These two fuels were fed separately into the engine, where they were mixed before combustion. Liquid oxygen is a common oxidizer used in rocket engines. Incidentally, in order for it to exist in liquid form, oxygen must be cooled below $-183°C$ ($-297°F$), and this is why ice can sometimes been seen on the exterior of a liquid-fuelled rocket on the pad, which then breaks away spectacularly in chunks during lift-off.

There are also 'mono-propellant' liquid engines – running on just a single tank of fuel. These, though, tend to be confined to smaller rockets and to the thrusters used to orient spacecraft once they have left the Earth. One such mono fuel is the chemical hydrazine, which when passed over a catalyst material breaks down into a high-temperature gas of hydrogen, nitrogen and ammonia to create thrust.

Straddling the gap between liquid and solid rockets is a final type of rocket known as a 'hybrid engine'. These use a solid fuel and a liquid oxidizer, making them less complex than a liquid engine yet more controllable than a pure solid rocket. SpaceShipOne, operated by

Mojave Aerospace Ventures, which in 2004 achieved the first crewed space launch by a private organization, was powered by a hybrid engine running on solid rubber as fuel with an oxidizer of nitrous oxide.

In all types of engine, the high-temperature, high-pressure gas resulting from combustion must be converted into a high-speed jet of exhaust. This is done using a rocket nozzle, a cone-shaped aperture that sits directly beneath the combustion chamber. One of the most efficient designs is known as the de Laval nozzle, after its co-inventor, Gustaf de Laval, who proposed it in 1888 for use in steam turbines. It consists of a tube pinched in the middle into a lop-sided hourglass shape, which constricts sharply on the inlet side, where the hot gas enters the nozzle, and then fans out into a more gently curved bell shape for the outgoing exhaust to push against. The de Laval nozzle can convert the high-pressure gas generated inside a typical rocket engine into a supersonic exhaust jet, travelling at thousands of metres per second.

One of the most high-performance rocket engines built to date is the liquid-fuelled Raptor engine, under development by SpaceX for its forthcoming super-heavy-lift Starship rocket. During a test firing in February 2019, the Raptor took the record for the highest recorded combustion chamber pressure, 3,900 psi (pounds per square inch) – that's about the weight of a large car concentrated onto a square just a couple of centimetres on each side.

MEGA ROCKETS

Name	Height (metres/ feet)	Payload to orbit (metric tonnes/ US tons)	Nation	Organization	First launched
Saturn V	110.6/362.9	140/154.3	USA	NASA	1967
N1	105/344.5	95/104.7	USSR	Energia	Four failed attempts 1969–72
Delta IV Heavy	72/236.2	28.8/31.7	USA	United Launch Alliance	2004
Angara-5	64/210	24.5/27	Russia	Khrunichev	2014
Falcon Heavy	70/229.7	63.8/70.3	USA	SpaceX	2018
SLS Block 1	98.1/321.9	95/104.7	USA	NASA	Planned 2020
Starship	118/387.1	150/165.3	USA	SpaceX	Planned 2021
SLS Block 2 Cargo	111.3/365.2	130/143.3	USA	NASA	Planned 2025–30
Yenisei	80/260 approx	70/77+	Russia	JSC SRC Progress	Planned 2028
Long March 9	93/305.1	140/154.3	China	CALT	Planned 2030

When development of the Raptor is complete, the final chamber pressure is anticipated to reach 4,400 psi, which the engine's de Laval nozzle will convert into an exhaust stream travelling at an astonishing 3,400 m/s (12,240 kph, or 7,606 mph), ten times the speed of sound in air.

Acting on impulse

Exhaust speed is one measure of a rocket engine's efficiency – that is, the fraction of the total chemical energy stored in the fuel that can eventually get converted into the motion of the rocket. The de Laval rocket nozzle raises the exhaust speed dramatically – making it supersonic (faster than the speed of sound), thereby increasing the efficiency from just a few per cent to, usually, in excess of 60 per cent. This is surprisingly high, given how noisy and inefficient rockets might seem.

Another measure of rocket efficiency that you may encounter, but which we won't dwell too much on here, is 'specific impulse'. This translates as the total amount of 'thrust' the engine delivers per unit mass of fuel burned. Thrust is the upward force pumped out by the rocket. It acts against the downward force, or weight, caused by the pull of gravity acting on the rocket's mass. If the rocket is to take off, then the thrust must exceed the weight.

It's quite possible to have an engine that is very inefficient (i.e. with low specific impulse and low exhaust speed), yet that still delivers sufficient thrust to take off (even if its low efficiency means it won't get far). And, conversely, there are extremely efficient engines that, while making excellent use of their fuel, do not have sufficient thrust to overcome the Earth's gravity and actually get off the ground.

One such rocket is known as an ion engine. Rather than using chemical energy released by combustion to create a high-speed exhaust jet, these rely on electric fields to accelerate charged fuel particles to extraordinary speeds, up to 50,000 metres per second (164,000 feet per second) – more than ten times higher than can be achieved with a conventional rocket. Yet the rate at which they eject these particles is so slow that their thrust is feeble and they are barely able to lift a few grams in Earth's gravity, so nowhere near their own weight.

> **When you're getting ready to launch into space, you're sitting on a big explosion waiting to happen.**
>
> SALLY RIDE
> astronaut and physicist (1988)

You might rightly wonder then, just what use they are. It turns out that an ion engine's high efficiency comes into its own in deep space, far away from the strong gravity of planetary bodies. Boosted into space by a conventional rocket, an ion-powered spacecraft spends its fuel slowly but wisely over weeks, months, even years – gradually accumulating a large delta-v from a relatively small mass of fuel, enabling it to travel great distances through the void of outer space.

In 1998, NASA's Deep Space 1 technology tester mission blasted off from Earth aboard a Delta II – a liquid-fuelled rocket, with three solid rocket boosters. Once in space, it fired up an ion engine, which took it on a journey across the solar system, flying past a comet and an asteroid, and returning pictures and scientific data from both. Over the course of its three-year mission, the spacecraft's ion engine changed its speed by more than 4,000 metres per second (13,120 feet per second), using less than 74 kg (163 lb) of its xenon gas fuel.

> **Holy flying fuck, that thing took off!**
>
> ELON MUSK (2018)

Scientists are now devising new engine technologies capable of achieving even higher exhaust speeds – hundreds of thousands or millions of metres per second. These rockets, essentially elaborations on Tsiolkovsky's original ideas more than a hundred years ago, could power the spacecraft that one day take human beings to the furthest reaches of the solar system and beyond.

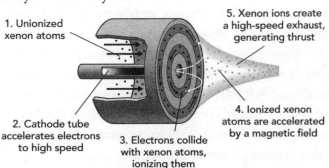

1. Unionized xenon atoms

2. Cathode tube accelerates electrons to high speed

3. Electrons collide with xenon atoms, ionizing them

4. Ionized xenon atoms are accelerated by a magnetic field

5. Xenon ions create a high-speed exhaust, generating thrust

03 USE THE FORCE

'The recommendation not to throw yourself out of a second-floor window is a part of the science of mutually gravitating bodies.'

ARTHUR BALFOUR (1893)

The landscape of the cosmos is sculpted by gravity. The gravitational field of the Earth fixes how the moon circles around it. Gravity has held the planets of our solar system turning in their clockwork orbits around the sun for the past 4.6 billion years. The sun itself formed from the collapse under gravity of a vast cloud of hydrogen gas, and now, like its attendant retinue of planets, it wheels around the centre of our Milky Way galaxy, doing so once every 230 million years. Even the Milky Way is part of a wider network of galaxy clusters and filaments of cosmic material that thread the universe, all held in place by gravity and dancing to its tune. From humble apples dropping off trees to the birth and death of the entire universe, the architect is always gravity – and gravity alone.

So it's probably no surprise that, just as the gravity of the sun determines the orbits of the planets, it also guides the trajectories of spacecraft winging their way across the solar system from one world to the next. We've seen that overcoming the Earth's gravity is a major obstacle to reaching space, and rockets must expend a vast amount of fuel to do this. However, once in space, gravity can be the space explorer's friend – if you know what you're doing.

The big G

The first scientific theory of gravity is credited to the British mathematician and physicist Sir Isaac Newton, who published it in 1687. His 'law of universal gravitation' gives the size of the gravitational force acting between two bodies, given their respective masses and the distance between them. The law provided an accurate description for the motion of the planets. Indeed, Newton used it to derive the three laws of planetary motion that had been formulated by the German astronomer Johannes Kepler earlier in the seventeenth century. Kepler arrived at his laws by poring over

> **Although gravity is by far the weakest force of nature, its insidious and cumulative action serves to determine the ultimate fate not only of individual astronomical objects but of the entire cosmos.**
>
> PAUL DAVIES (1994)

tables of astronomical observations, logging how the positions of the planets changed with time, and looking for underlying relationships that could explain the data. Newton's law of gravity gave Kepler's empirical model a solid scientific basis.

KATHERINE JOHNSON (1918-2020)

Creola Katherine Coleman (Johnson was the surname of her second husband) was born on 26 August 1918 in White Sulphur Springs, West Virginia. She showed natural talent for mathematics and graduated with the highest distinction possible from West Virginia State College in 1937, at the age of just eighteen. Later she became only the third African-American to be awarded a mathematics PhD. After raising a family with her first husband, in 1953 she joined NACA (the National Advisory Committee for Aeronautics, which later became NASA). Her role was a 'human computer' – manually carrying out complex calculations that today would be done electronically. She famously calculated the trajectory for the flight that made Alan Shepard the first American in space. As a woman of colour in a field dominated at the time by white men, she had to overcome considerable adversity throughout her career. In 2015, she was presented with the Presidential Medal of Freedom.

Kepler's laws show that the orbits of the planets don't have to be circular but, in general, follow elliptical (oval-shaped) paths. Most of the planets of the solar system do, in fact, follow roughly circular paths around the sun, the only notable exception being the innermost world, Mercury, whose orbit brings it to a closest approach to the sun (known as 'perihelion') of 46 million km (29 million miles) and carries it as far away (a point known as 'aphelion') as 70 million km (43 million miles).

The same laws apply to spacecraft that are coasting through the solar system – that is, not firing their engines but moving purely under the guidance of gravity. At the time of writing, both NASA's Parker Solar Probe and the Chinese Chang'e 2 mission are on planet-like solar orbits – they are effectively artificial satellites of the sun. The Voyager spacecraft, probes sent to the outer planets of the solar system, launched in the 1970s, are also cruising freely in the gravitational field of our star, although these craft have now entered interstellar space, and are travelling so fast that not even the might of the sun's gravity will ever bring them back home.

But we're jumping the gun. In the last chapter, we saw how rockets are used to boost spacecraft from the surface of the Earth, up to the Kármán line, 100 km (62 miles) up, which marks the boundary between the planet's atmosphere and outer space. Given what Newton and Kepler have told us about gravity and orbits, how exactly

THE RULES OF THE SOLAR SYSTEM

German astronomer Johannes Kepler deduced his laws of planetary motion, which actually govern everything in the solar system including comets, meteoroids – and cruising spacecraft – after studying data gathered by the Danish observational astronomer Tycho Brahe.

We won't go into the gory mathematical details but here, in essence, is what they say. The first law holds that planets move in elliptical (oval-shaped) orbits, with the sun at one of the 'focal points' (the ellipse equivalent to the centre of a circle). Law two says that a line drawn from the sun to a planet sweeps out equal areas in equal times as the planet moves. And the third and final law states that if you square the time taken by a planet to complete an orbit then this number is proportional to the cube of the orbit's size. So if the size of the ellipse quadruples, the orbital time increases by a factor of eight (4 cubed = 64, square rooted = 8).

does a spacecraft go from the shores of the cosmic ocean to charting a course across interplanetary space?

Casting off

The simplest kind of trip into space that gravity permits is what's known as a 'sub-orbital' flight – the rocket blasts off from the Earth's surface and flies in a giant arc, carrying it briefly up above the Kármán line, before the planet's gravity pulls it back down to the ground. When Alan Shepard became the first American in space on 5 May 1961, this is exactly the path that he followed on his fifteen-minute flight. Shepard's capsule launched from Cape Canaveral, Florida, atop a Mercury-Redstone rocket and reached a maximum altitude just shy of 188 km (117 miles) before splashing down in the Atlantic Ocean 487 km (303 miles) downrange.

The space tourism operator Virgin Galactic is soon to start taking passengers on short sub-orbital excursions into space, boosting them above the Kármán line in a rocket-powered spaceplane and then landing back on a runway like an ordinary aircraft. And sub-orbital trajectories are still used on some scientific research flights, where small solid-fuelled 'sounding rockets' propel experiment payloads into space to briefly gather data before falling back to the ground.

> **We didn't slow down, unlike the others, when we got to the moon because we needed its gravity to get back.**
>
> JIM LOVELL
> commander, Apollo 13 (2011)

The next destination beyond the Kármán line is Earth orbit. The handy thing about orbits, of any sort, is that they are self-sustaining. That is, as long as the spacecraft is high enough to be clear of any drag from the atmosphere of the planet below, which might slow it down, it will continue to orbit indefinitely with no need to fire its engines.

To see how, imagine a cannon perched atop a high mountain. The cannon shoots projectiles horizontally off the mountain top, and each shot packs a little more punch than the one before. The first shot plops out of the cannon and lands at the foot of the mountain, but as the power increases they land further and further away until, eventually, gravity cannot pull the projectile down to the ground and it loops right round the planet. The projectile is still falling under gravity but it's moving so fast that the surface of the planet, due to its curvature, is falling away at the same rate.

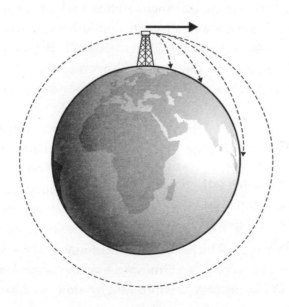

Low Earth orbit (LEO) begins at around 200 km (124 miles) above the surface of the Earth and extends up to 2,000 km (1,243 miles). The lowest LEO path requires a speed of around 7,800 m/s (28,000 kph, or 17,400 mph) to maintain, so spacecraft in such orbits typically zip around the Earth once every ninety minutes. The permanently crewed International Space Station orbits the Earth in LEO at an altitude of just over 400 km (250 miles). LEO is usually the jumping-off point for space missions destined to travel further afield, sometimes referred to as a 'parking orbit', and so is the first port of call after leaving Earth.

One consequence of Kepler's laws is that the higher a spacecraft orbits above a planet, the slower it moves. For low Earth orbit at an altitude of 1,000 km (620 miles), the orbital speed drops down to 7,400 m/s (26,640 kph, or 16,550 mph), lengthening the time to complete one orbit from 90 to 105 minutes. Keep increasing the altitude like this and eventually the duration of one orbit will equal twenty-four hours, the rotation period of the Earth. This occurs at an altitude of 35,786 km (22,236 miles).

If this orbit is also directly over the Earth's equator then, to someone on the ground looking up, the spacecraft will appear to hang in the sky, as it circles in lockstep with the Earth's rotation. This is known as a 'geostationary' orbit.

Communications and TV satellites are placed in such orbits so that receiver dishes on the ground don't have to

> **Get to low Earth orbit, and you're halfway to anywhere in the solar system.**

ROBERT HEINLEIN (1950)

move and track the satellite in order to receive the signal. The requirement that the satellite orbits over the equator is why, in the northern hemisphere at least, satellite TV dishes on houses always point in a southerly direction.

Other types of orbit include 'polar', where the orbit is at 90 degrees to the equator so that the satellite passes over the poles; and 'sun synchronous', where the satellite always lies between the sun and the planet, often so that it can observe the planet's surface in daylight.

Getting from A to B

For the most exciting space missions, low Earth orbit is just the beginning. Mission controllers will pick their moment and then fire the spacecraft's engines to move it out of LEO and onto a so-called 'transfer' orbit. First conceived by the German engineer Walter Hohmann in 1925, this is an elliptical trajectory bridging the gap between two roughly circular orbits. On arrival, a second engine burn takes the spacecraft off the transfer trajectory and onto its final orbit.

A transfer orbit might bridge the gap between two orbits at different altitudes above the Earth. Or, it could involve hopping from the Earth's orbit around the sun to that of another planet in the solar system. In this case, careful planning is essential – it's one thing simply to jump between orbits, but when transferring from planet to planet the spacecraft's arrival must be timed to coincide with the passage of the destination world. For example, it's no use reaching the orbit of Mars if the planet itself is on the opposite side of the sun.

This gives rise to the notion of 'launch windows' – tight intervals of time during which an interplanetary space mission to a given destination can depart the Earth. For example, NASA's Curiosity Mars rover mission had a launch window from 25 November 2011 to 18 December 2011, and even then, the daily rotation of the Earth – which constantly changed the direction in which the rocket would leave the planet – meant that the window was open for less than two hours each day. Happily, the mission blasted off successfully on 26 November, landed on 6 August 2012, and as of late March 2020 is still busy exploring the surface of the Red Planet.

> **Everything was so new – the whole idea of going into space was new and daring. There were no textbooks, so we had to write them.**
>
> KATHERINE JOHNSON
> NASA mathematician (1999)

Interplanetary transfers come with added complications, such as having to account for the gravity of the Earth when firing the engines to leave and that of the destination planet on arrival. When the Apollo astronauts went to the moon, the transfer trajectory was especially difficult because, given the proximity of the Earth and the moon, the gravity of both bodies was pulling on the spacecraft simultaneously. The solution turned out to be a figure-eight loop, taking the Apollo spacecraft around the back of the moon and then onto a return path to Earth.

This is a classic example of a 'three-body' gravitational problem, and they are notoriously tough to solve mathematically. One particular instance of the three-body problem was solved by the Italian mathematician Joseph Louis-Lagrange in 1772. He found five points in space in a two-body gravitational system where a third body can be added so that it maintains its position relative to the other two. These so-called 'Lagrange points' are labelled 'L1' to 'L5' and form a cross shape. For example, L1 is a kind of equilibrium point where the gravity of the first two bodies in some sense balances out. An object placed at L1 in the Earth-sun system will circle around the sun in step with the Earth.

The Lagrange points are a useful parking spot for spacecraft. A number of sun-observing probes have been stationed at the Earth-sun L1 point, while the James Webb Space Telescope, the successor to the Hubble Space Telescope, due to be launched in 2021, will sit at L2, on the far side of the Earth from the sun, from where it will be able to observe the heavens in perpetual darkness.

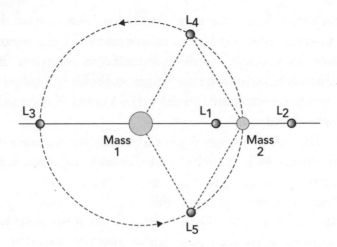

Even while still travelling to your destination, gravity can be your friend. That's because the gravitational field of other planets can sometimes be exploited to boost the speed of a cruising spacecraft without the need to fire its engines. Manoeuvres such as this are known as 'gravity assists', and are a trick used by wily mission planners to save fuel when aiming for far-flung destinations.

A spacecraft approaching a stationary planet experiences a gravitational pull accelerating it and then, once it's flown past, an equal and opposite force slowing it down again. But planets are not stationary – as we've seen, their orbits keep them in constant motion around the sun – and a moving planet transfers some of its momentum to the spacecraft as it passes. It's a bit like bouncing a ping-pong ball off a bat. If the bat is stationary then the ball bounces back with much the same speed that it came in with. However, swing the bat vigorously and the ball emerges from the collision travelling much faster. It's much the same with gravity – a moving planet is able to haul in a passing spacecraft and then slingshot it onwards at a much higher speed.

The NASA Voyager 2 mission to the outer planets of the solar system, launched in 1977, relied on a string of gravity-assist manoeuvres to hop from one world to the next. This was made possible by a rare alignment of the planets, which only takes place once every 176 years. Gravity assists can also be configured to slow a spacecraft down if need be. And this has been done on missions to Mercury and Venus, the inner worlds of the solar system, where the sun's gravity tends to accelerate the spacecraft as it travels inwards, creating excess speed that needs to be shed.

> **In some sense, gravity does not exist; what moves the planets and the stars is the distortion of space and time.**
>
> MICHIO KAKU (2000)

Everything's relative

While Newton's law of gravity does an astoundingly good job of modelling the behaviour of spacecraft plying the solar system, it wasn't the last word from scientists about how the gravitational force actually works. In 1915, German physicist Albert Einstein published his general theory of relativity, which attributed gravity to curvature of space and time. In this picture, massive celestial bodies distort the space-time landscape, creating hills and valleys, which then in turn dictate the motion of objects that are travelling through it. In our solar system, the mass of the sun creates a giant bowl-shaped depression in space, which the planets are effectively rolling around the inside of as they move along their orbits.

Einstein had been led to the general theory after formulating his 'special' theory of relativity ten years earlier. Special relativity was a new take on the science of moving bodies, which, along the way, had given rise to the famous equation $E=mc^2$, linking energy (E) and mass (m) by the speed of light (c), and which forms the basis for nuclear energy. But the special theory took no account of gravity. To try and remedy this, Einstein undertook thought experiments in which he constructed scenarios in special relativity, then imagined them falling in a gravitational field, and tried to reason out in his mind what should happen. These mental games convinced him that the correct way to account for gravity was to bend the flat space and time of his special theory. And the resulting general theory of relativity has since been verified experimentally.

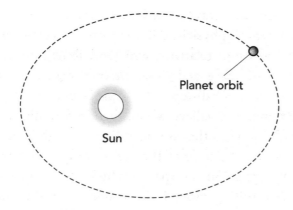

Planet orbit

Sun

What Newtonian gravity lacks in accuracy, it makes up for with its simplicity. General relativity is a complex theory and, in most cases, Newton's law is a good enough approximation for analysing spaceflight within the confines of the solar system.

04 HOW TO STAY ALIVE IN SPACE

'Houston, we've had a problem.'

JIM LOVELL, COMMANDER, APOLLO 13 (1970)

Dozens of brave souls have perished in the quest to explore the high frontier. The essence of the problem is simple enough: human beings are delicate creatures and space travel is brutal. Getting accelerated to twenty-five times the speed of sound, starved of oxygen, bathed in lethal radiation, heated to in excess of 1,600 °C (2,900 °F) and dropped back to Earth from hundreds of kilometres up isn't without its risks.

We can blame evolution. Our species has emerged and adapted to a life of relative comfort here on Earth. None of the challenges thrown up by our planet's environment could ever prepare us for the deadly extremes of outer space. Instead, astronauts must look to engineers to provide some safety. And that's no mean feat. Indeed, keeping human travellers alive in the teeth of outer space presents just as much of a challenge as getting them there in the first place.

A deadly game

In fact, the danger begins way before you even leave Earth. This was chillingly demonstrated on the morning of 28 January 1986, when the world watched in horror as the Space Shuttle *Challenger* disintegrated in mid-air, just 73 seconds after lift-off from Cape Canaveral, Florida, killing all seven crewmembers. These included Christa McAuliffe, a civilian who NASA had selected to become the first teacher in space, and Ronald McNair, a physicist turned astronaut, and one of the first African-Americans to go into space.

The fault was ultimately traced to the shuttle's solid rocket boosters. These are made up of seven cylindrical steel sections, which are loaded with solid propellant and then clamped together, each joint sealed with a pair of washer-like rubber O-rings. The O-rings were tested by their manufacturer, Thiokol, down to a temperature of 4 °C (39 °F), but the morning of the launch was exceptionally cold, −1 °C (30 °F), having dropped to −8 °C (18 °F) overnight. The cold made the O-rings brittle and, at 59 seconds into the flight, a plume of flame erupted from the lower joint of the right-hand booster rocket. The plume went to work like a blowtorch on the side of the shuttle's external fuel tank, which, 14 seconds later, exploded catastrophically.

The shuttle wasn't destroyed by the explosion itself. In a split second, it was knocked onto a tumbling trajectory by

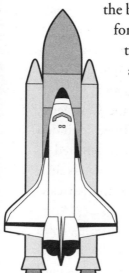

the blast, subjecting it to extreme aerodynamic forces (the craft was already supersonic by this time) that tore its lightweight aluminium airframe apart. The crew cabin can be seen in photographs emerging from the fireball intact. It took two minutes and 45 seconds to fall the 15 km (9 miles) back to Earth, hitting the surface of the Atlantic Ocean at 330 kph (200 mph).

When the wreckage of the cabin, along with the remains of the astronauts, was recovered from the seabed about 29 km (18 miles) north-east of Cape Canaveral, emergency air packs were found to have been activated – proving that at least some of the crew were alive and conscious after the break-up of the spacecraft.

The *Challenger* disaster is a stark reminder of what a deadly game space travel really is. But it's also a cautionary tale about the pitfalls of cutting corners – Thiokol had forewarned NASA that the O-rings were not certified for use in such low temperatures. Yet, despite that advice, NASA still opted to proceed.

Tragedy might have been averted had the shuttle been fitted with some kind of crew escape system. For its initial four flights, the astronauts did have ejector seats. This was feasible, since these early test missions

> ❝ **If we die we want people to accept it. We are in a risky business ... The conquest of space is worth the risk of life.** ❞
>
> VIRGIL 'GUS' GRISSOM
> (1966)

operated with just a two-person crew. However, once the shuttle entered operational service, the crew size began to increase, up to its maximum of seven. Four of these crewmembers were seated on the flight deck during launch, while the other three were on another deck directly below – making it impossible to eject. After *Challenger*, NASA did add a parachute escape system that could be deployed during the shuttle's glide back to Earth, but this would not have been usable during the shuttle's powered ascent.

Bail out!

Modern crewed launch vehicles, and many of those pre-dating the shuttle, utilize what's called a 'launch escape system'. On the Saturn V, for example, this took the form of a small rocket mounted directly above the crew capsule. In the event of an emergency, the capsule separates from the main launch vehicle and the escape rocket fires, pulling the crew out of danger. Parachutes then slow the capsule's descent to the ground.

Launch escape systems have saved lives. In 1983, the Russian Soyuz 7K-ST mission burst into flames on the

APOLLO 13

Perhaps the greatest tale of triumph over the adversities posed by spaceflight is that of the Apollo 13 mission to the moon. Intended to be the third crewed lunar landing, the mission launched from Cape Canaveral on 11 April 1970. At around fifty-six hours into the flight, with the spacecraft two-thirds of the way to the moon, an electrical fault caused one of the onboard oxygen tanks to explode.

The blast punctured a second tank, and shut off two of the three oxygen-hydrogen fuel cells used to generate electrical power. With oxygen venting into space, and only one-third power, the challenge now was to get the astronauts home alive. Through a combination of ingenuity, and cannibalizing batteries, oxygen and other equipment from the lunar landing module (fortunately, they were still to reach the moon when the accident occurred), the command module splashed down safely in the Pacific Ocean on 17 April. Astronauts Fred Haise, Jack Swigert and Jim Lovell, as well as the mission's ground control team at Houston, all received the Presidential Medal of Freedom.

launch pad. The escape rocket fired, hauling the crew capsule clear seconds before the Soyuz exploded (there's video online – google it!). Cosmonauts Vladimir Titov and Gennady Strekalov, though badly bruised by the escape rocket's violent acceleration, lived to tell the tale.

Escape systems are effective on rockets whose stages are stacked up one on top of the other (known as 'serial staging' – see Chapter 2). In this configuration, the crew capsule sits at the top of the stack, making it easier to pull clear if anything goes wrong. This system probably wouldn't have helped the Space Shuttle, which is a parallel stack – the crew are situated alongside the main fuel tank, making it inherently more dangerous.

The good news is that most rocket launches make it to space intact. But that's just the beginning. The first issue for a human being suddenly deposited in the space environment is the lack of any air to breathe. The tiny traces of gas present in space tend to get drawn in by gravitating bodies, which is why planets (at least, the bigger ones) have atmospheres, while space itself is a hard vacuum.

> **I could feel the saliva on my tongue starting to bubble just before I went unconscious.**
>
> JIM LEBLANC (2010)

There's no real data on what would happen to a human being subjected to a vacuum for an extended period. It's been conjectured that the body would swell up, fluids would boil and, if you believe Hollywood, the eyes pop out. The best information we have is the case of engineer Jim LeBlanc who, in 1966, was testing a prototype spacesuit in a NASA vacuum chamber when the hose pressurizing his suit became disconnected. He passed out after 14 seconds, but recalls feeling the saliva boiling on his tongue due to the low pressure. He was given emergency oxygen seconds later, and made a full recovery.

Spacesuits serve as direct protection from the vacuum of space for astronauts while out on spacewalks and as an emergency backup during normal operations. Most of the time the crew are in the relative safety of their spacecraft, which is pressurized and generates its own oxygen from a number of sources. Short-duration spaceflights simply bring tanks of oxygen with them. On the International Space Station (ISS) the gas is manufactured by passing an electric current through water to separate it into hydrogen and oxygen, the unwanted hydrogen being vented to space. The air on the space station is pumped through filters to remove odours and impurities, and it's passed over cold metal plates to extract moisture. Devices called 'CO$_2$ scrubbers' soak up harmful carbon dioxide, unavoidably produced as the astronauts exhale. They work by reacting the gas with lithium hydroxide to turn it into lithium carbonate and water.

Oxygen generators are not infallible. In 1997, the system in use on the Soviet Mir space station was destroyed by fire. The fire was a major incident in itself, but in its wake the crew, faced with the prospect of having no breathable air, resorted to emergency lithium perchlorate candles, which give off oxygen as they burn. Canisters of lithium perchlorate are still carried as a backup on the ISS today.

Hypersonic shrapnel

Maintaining the divide between the harsh vacuum outside and breathable atmosphere within is a big challenge for spacecraft designers. The greatest risk comes from collisions with small particles that, despite their size, travel so fast that they are capable of punching a hole in a spacecraft's outer skin. At best, this will cause a small leak that can be fixed. At worst, it can lead to explosive decompression – the spacecraft's hull popping like a balloon as the air inside rushes out.

Some of these particles will be natural micrometeoroids (tiny particles of rock), but, amazingly, most encountered in Earth orbit consist of manmade space debris. They range from bits of spent rocket stages, to dropped tools, fragments of defunct satellites, and even frozen urine. Travelling at nearly 8 km/s (5 miles/s), a 5-gram bolt packs as much energy as a 200-kg weight dropped from the top of an eighteen-storey building.

At this speed, even flecks of paint can strike like a blizzard of hypersonic shrapnel. And it's happened. In 1983, a tiny speck left a significant pit in the windshield of Space Shuttle

> **Spaceflight will never tolerate carelessness, incapacity, and neglect.**
>
> GENE KRANZ
> NASA flight director (1967)

Challenger. In 2007, a piece of space debris punched a 6-mm hole clean through one of Space Shuttle *Endeavour*'s radiator panels.

The European Space Agency (ESA) estimates there are some 900,000 pieces of space debris larger than a centimetre in size, and many millions smaller than this. The US Space Surveillance Network (SSN), a global array of tracking stations, is currently monitoring some 20,000 objects in orbit around the Earth – only around two thousand of which are operational spacecraft. When the SSN detects an object due to pass too close to an active spacecraft (usually determined as the probability of an impact exceeding one in ten thousand) it alerts controllers so that the craft can take evasive manoeuvres. The ISS has to do this, on average, once per year.

To be trackable by the SSN, a piece of debris must be at least 10 cm in size. To deal with smaller pieces that can't be dodged, the ISS is protected with what are called Whipple shields, after the US astronomer Fred Whipple, who came up with the idea. These consist of spaced-out layers of shielding to shatter an inbound piece of space junk and spread the fragments over a wider area. Diffusing the impact in this way – effectively turning a rifle bullet into a shotgun blast – makes it easier for the spacecraft's hull to absorb it. In the case of the ISS, the hull is made from aluminium, reinforced with layers of the ceramic fabric Nextel, as well as Kevlar (also used in bulletproof vests).

The phantom menace

Bits of debris aren't the only things astronauts get pelted with. There's also an invisible menace to contend with: radiation. It's a major hazard in space, especially on long-duration flights, and is perhaps the single most problematic factor standing in the way of plans to send human explorers to the other worlds of our solar system.

Most of the harmful radiation in deep space takes the form of high-speed subatomic particles. These are either spewed out by the sun during violent events on its surface such as solar flares and coronal mass ejections, or they are cosmic rays, ultra-high-energy particles that originate from beyond the solar system – perhaps even from beyond our own galaxy. Supernova explosions, marking the death of very large stars, have been identified as one possible source.

Most of these radiation particles are electrically charged, which is good news for us down here on Earth as they get batted away by the planet's magnetic field. And those that do get through are largely absorbed by the atmosphere. It's a different matter up in space though, and astronauts are especially vulnerable.

Apollo crewmembers reported seeing occasional bright flashes, which are believed to have been radiation particles passing through their eyeballs. The Apollo programme actually dodged a bullet. A violent solar storm, powerful enough to disrupt electrical grids on Earth, erupted in August 1972 – right between the Apollo 16 and 17 flights. Had it occurred during one of these missions it could have meant death, or severe radiation sickness, for a moonwalking astronaut.

> **One day in space is equivalent to the radiation received on Earth for a whole year.**
>
> MARCO DURANTE
> physicist (2019)

On a trip to Mars, it's estimated a human could absorb up to seven hundred times the ambient dose of radiation received by someone remaining back on Earth. There's also evidence that astronauts show higher risk for developing cataracts and heart disease later in life, both ascribed to radiation exposure. Add to that the enhanced risk of cancer and nervous system damage on extended flights.

CASUALTIES OF THE SPACE PROGRAMME

Date	Mission	Nation	Description	Fatalities
23 Mar 1961	Cosmonaut training	USSR	Fire in a low-pressure altitude chamber	Valentin Bondarenko
27 Jan 1967	Apollo 1	USA	Electrical fault ignited pure oxygen atmosphere in cockpit during pre-launch test – crew burned to death	Virgil 'Gus' Grissom, Ed White, Roger B. Chaffee
24 Apr 1967	Soyuz 1	USSR	Parachute failed to deploy after re-entry, causing capsule to impact the ground at high speed	Vladimir Komarov
15 Nov 1967	X-15 Flight 3-65-97	USA	Electrical fault caused pilot to lose control while travelling at Mach 5, causing craft to break up in flight	Michael J. Adams
8 Dec 1967	Astronaut training flight	USA	F-104 Starfighter jet crashed rehearsing rapid-descent glides, required to land a space plane	Robert H. Lawrence Jr
30 Jun 1971	Soyuz 11	USSR	Faulty valve vented all oxygen from capsule after undocking from Salyut 1 space station	Georgy Dobrovolsky, Viktor Patsayev, Vladislav Volkov

Date	Mission	Nation	Description	Fatalities
28 Jan 1986	STS-51-L	USA	Space Shuttle *Challenger* is destroyed 73 seconds into flight, after cold weather causes solid rocket booster to fail	Gregory Jarvis, Christa McAuliffe, Ronald McNair, Ellison Onizuka, Judith Resnick, Michael J. Smith, Dick Scobee
11 Jul 1993	Cosmonaut training	USSR	Drowned during water recovery training in the Black Sea	Sergei Vozovikov
1 Feb 2003	STS-107	USA	Space Shuttle *Columbia* burns up on re-entry due to damage sustained by heat shield during launch	Rick D. Husband, William C. McCool, Michael P. Anderson, David M. Brown, Kalpana Chawla, Laurel Clark, Ilan Ramon
31 Oct 2014	Virgin Galactic test flight	USA	Virgin's SpaceShipTwo VSS *Enterprise* breaks up when descent wings accidentally deployed during powered flight	Michael Alsbury

> **The true courage of space flight is not sitting aboard 6 million pounds of fire and thunder as one rockets away from this planet. True courage comes in enduring... persevering, the preparation and believing in oneself.**
>
> RONALD MCNAIR
> Challenger Space Shuttle
> astronaut (1984)

The most straightforward way to limit radiation exposure in space is regular crew rotations. This is a simple matter on the Earth-orbiting International Space Station, where the typical stay for an individual astronaut is about six months. Even this would be excessive in deep space, but the ISS is afforded some protection as its low orbit keeps it inside the natural shield afforded by Earth's magnetic field, which deflects away charged radiation particles. But a crewed mission to Mars would mean many months exposed in deep space, with no option to get off.

Of course shielding can block radiation, but shielding is heavy and, as we've seen, when it comes to launching anything into space, weight means more fuel, which is money. The compromise on the ISS is a single, heavily shielded module that the crew can all pile into during times of intense solar activity. Drugs are another possibility – which slow the rate at which cancer-causing DNA damage can spread, giving cells time to heal.

One innovative idea under development is to generate

an artificial magnetic field around a spacecraft, to deflect away radiation particles much like the Earth's magnetic field. It used to be thought that for this to work, the magnetic bubble surrounding a spacecraft would have to be many kilometres across and require megawatts of power to generate. But researchers at Rutherford Appleton Laboratory, in the UK, have found – through experiment and computer calculations – that a far more modest field, just 100 metres (330 feet) across, would be sufficient. They are currently working with NASA and other institutions to develop the concept, which could turn out to be a crucial enabling technology for long-haul human spaceflight.

The voyage home

Most humans who travel into space will do so on a return ticket, which means landing them safely back on Earth. On a sub-orbital flight that's not too difficult. There are two main methods. The first is to deploy parachutes – either coming down on water or, with the help of rockets or airbags to cushion the impact, solid ground. The other option is to glide down on wings and land on a runway, much like an aircraft. And both these techniques have been successfully demonstrated.

But far more perilous is making the return journey from orbit. The problem is the speed involved – reaching low Earth orbit means accelerating to 7.8 km/s (4.8 miles per second), equivalent to 28,000 kph (17,450 mph). And to land back on Earth a spacecraft must shed all this. You might think the simplest way to slow down is to fire retro rockets. But braking from such a speed requires a lot of fuel, all of which would need to be hauled up into orbit, demanding a *huge* amount more fuel when the spacecraft launched and an unfeasibly large and expensive rocket.

Instead, mission planners opt to use the Earth's atmosphere as a natural braking system. The spacecraft fires its rockets briefly to begin its descent from orbit. As it descends, the atmosphere gradually becomes denser, creating a drag force (a resistance to motion as a body moves through a fluid) that slows the craft down. The issue is that this generates a lot of heat. The air in front of the spacecraft gets compressed and, just as a bicycle pump gets hot as you squash the air inside, so the air in front of the spacecraft heats up too – only, in this case, reaching temperatures of up to 1,600 °C (2,912 °F), hot enough to melt steel.

This is why spacecraft returning from orbit must be fitted with what's called a 'thermal protection system' – a heat shield to hold the searing temperatures of re-entry at bay. Spacecraft of the 1960s and 70s, including the Apollo missions, used so-called 'ablative' heat shields that would char as they heated up, allowing small pieces to break off and carry heat away. By their nature, ablative heat shields can only be used once.

The Space Shuttle, on the other hand, relied on a mix of lightweight ceramic tiles, carbon composites and insulation blankets. The shuttle's system was reusable, but there were concerns from the start over its fragility. These concerns were tragically validated on 1 February 2003, when the Space Shuttle *Columbia* burned up and disintegrated on re-entry. All seven astronauts were killed. The subsequent investigation found that a piece of insulating foam, which had broken away from the shuttle's external tank during launch, had punched a hole through the heat shield on the leading edge of the left wing. Upon re-entry, this allowed in superheated gas, which quickly melted the internal aluminium structure, causing the craft to break up.

> **If we want to go to Mars, it will be very, very difficult, it will cost a great deal of money, and it may cost human lives.**
>
> SCOTT KELLY
> astronaut (2017)

Perhaps because of the *Columbia* accident, the next generation of crewed spacecraft has returned to the tried and tested ablative heat shield design. And these craft are set to launch on rockets with the crew capsule at the very top – making it impossible for falling debris to damage the thermal protection system as it could, and did, on the shuttle.

Although the way is fraught with danger for human space travellers, innovation continues to overcome many of these hazards. And yet, there will always be destinations too perilous and missions that don't always demand the presence of an explorer made from flesh and blood. Which is why, as we're about to discover, it's sometimes best not to send humans at all.

05 WHEN TO LET THE MACHINES TAKE OVER

'Robotic exploration of the planets and their satellites as well as of comets and asteroids has truly revolutionized our knowledge of the solar system.'

JAMES VAN ALLEN (2004)

Mention robots in space and, for many of us, it will conjure up images of R2-D2 and C-3PO from *Star Wars*, or the wavy-armed mechanical assistant from *Lost in Space*. While the reality may be very different from these fictional portrayals, robots play an essential role in humanity's exploration of outer space. In fact, in every corner of the solar system where humans have left their mark (at least, beyond the moon), we've done so vicariously through our robotic emissaries.

They have the resilience to survive in extreme environments where fragile human astronauts wouldn't stand a chance. This is why when it comes to exploring the icy reaches of Pluto, flying through the searing heat of the sun's corona, or surviving the crushing pressures on the surface of Venus, robots are for now the only realistic choice.

Space travel's not just dangerous – it's also very expensive. Much of the cost of sending humans into space derives from the bulky life-support systems that they have to take with them. Making breathable oxygen, keeping a supply of food and water, and shielding the spacecraft to keep out harmful radiation, all add weight to a spacecraft, which thus requires more fuel and a bigger, more costly rocket to loft it into space.

> **Robots would do a much better job and be much cheaper because you don't have to bring them back.**
>
> STEPHEN HAWKING (2004)

Just for clarity, in the discussion that follows, I'm going to call any craft that went into space without a human crew 'robotic'. So, even if it was remotely piloted by humans on the ground, as far as we're concerned here, it's a robot.

Moon machines

The first robotic spacecraft was flown by the USSR in 1951, and carried two dogs, Dezik and Tsygan. In 1957, the Soviets launched Sputnik 1, the first craft to orbit the

THE AI REVOLUTION

As robot space probes travel further afield, it becomes more difficult for humans on Earth to control them. Even for a spacecraft at Jupiter, radio signals travelling at the speed of light can take in excess of forty minutes to travel each way. That makes real-time communication impossible, and complex tasks painfully slow. Artificial intelligence, or AI, is a field of data science that gives computers – and, by association, the robots that they control – a degree of autonomy, to make their own decisions and perform tasks unaided. These systems might help an orbiting satellite decide what kind of terrain it's flying over. Or it might allow a rover vehicle to drive itself automatically to a specified destination, selecting the optimal route. On a dynamically active planet, automated detection and avoidance of real-time hazards might be essential for a probe's survival. NASA's Curiosity Mars rover already uses an AI system to automatically identify which nearby rocks are interesting and deserve further analysis. And a raft of AI technologies are due to be deployed on forthcoming missions.

Earth, and then Sputnik 2, which carried another dog, Laika, on an orbital trajectory around the planet.

The US followed suit in 1958 with its own robotic orbiters, Explorer 1 and Vanguard 1. And it wasn't long before mission planners were sending uncrewed vehicles further afield. In 1959, the USSR began its Luna programme of robotic orbiter and lander missions to the moon. Luna 2 became the first human-made object on the moon's surface when it crash-landed there in September 1958. A year later, Luna 3 swung around our natural satellite to return the first ever images of its far side. The later Luna missions even collected soil samples from the moon and returned them to Earth. To date, these are the only successful robotic sample-return missions to a planet or moon.

> I hope that by 2050 the entire solar system will have been explored and mapped by flotillas of tiny robotic craft.
>
> MARTIN REES (2009)

In 1961, the US began its own robotic exploration of the moon, in the form of the Ranger missions, the first of which reached the lunar surface in April 1962. The aim of the Ranger programme was to obtain high-resolution images of the moon by placing the spacecraft on a collision course and returning pictures right up until the moment of impact.

Ranger was followed by the Surveyor programme,

also operated by US space agency NASA, but this time to make soft landings on the moon in preparation for the Apollo crewed landings. In particular, some scientists had warned that the lunar soil could be so light and powdery that any landing craft would simply sink down into it. The Surveyors proved that this is not the case. Surveyor 1 touched down successfully near Flamsteed Crater on 2 June 1966. In fact, in November 1969 Apollo 12 landed just 183 metres (600 feet) from Surveyor 3, allowing astronauts to walk to the probe and recover its TV camera and other parts. It remains the only robotic spacecraft to have been called on by humans at its landing site.

At the same time as Surveyor, the Soviet Union was developing its Zond programme. This had initially been intended to make flybys of Mars and Venus, but when the first two missions failed it was repurposed for the slightly less ambitious task of lunar exploration.

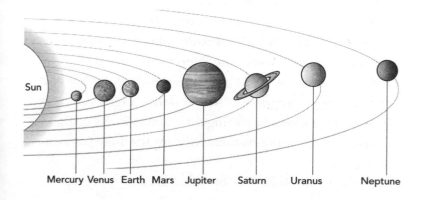

Mercury Venus Earth Mars Jupiter Saturn Uranus Neptune

Sun

Russia's plans for robotic exploration of the planets were far from dead, though. Its Venera programme to Venus fared very much better than Zond. Venera 1 flew past Venus in May 1961. It was the first ever flyby of another planet by a spacecraft from Earth. Sadly, contact with the probe was lost prior to this, so there were no pictures or data returned to Earth. In 1966, Venera 3 crash-landed on Venus, becoming the first spacecraft to make it down to the surface of another world. It was followed by Veneras 4–6, which made partial controlled descents while still transmitting data. Finally, on 15 December 1970, Venera 7 made the first soft landing on the planet, and survived on the 460 °C (860 °F) surface for 23 minutes. The programme continued into the 1980s, and in 1982, Venera 13 sent back the first colour images of the planet's hellish baked surface.

Red Planet

The United States also sent robotic craft to Venus in the 1960s and 70s. The first ever flyby of a planet with a functioning craft took place in December 1962, when NASA's Mariner 2 sped by Venus. But it was Mars where the US really led the way. Its Mariner 4 mission made the first flyby of the Red Planet in 1964. And on 14 November 1971, Mariner 9 became the first probe to orbit Mars, returning close-up images of the surface – including the giant rift valley scoring the planet's equator, which was named Valles Marineris in honour of the probe.

The US followed up the success of Mariner with the Viking Mars programme – two orbiters and two landers, which arrived successfully in July and September 1976. The missions revealed that much of the surface of Mars seems to have been shaped by water erosion. The landers returned stunning imagery and scientific data from the planet's surface.

Through the 1980s and 1990s, exploration of Mars seemed to go through a quiet period. And of those missions that did launch, many failed. Indeed, over the years Mars has gained a reputation as a notoriously difficult planet to get to, with fewer than half of the probes sent there achieving their full objectives. The run of bad luck began to change in the late 1990s, when NASA landed Mars Pathfinder on Ares Vallis, an ancient Martian flood plain. The mission was novel because it carried Sojourner, a small rover about the size of a microwave oven, which could drive around to investigate any interesting objects it found in the vicinity of the landing site. The USSR had actually pioneered the concept back in the 1970s, with its successful Lunokhod rover missions to the moon. Russia tried to send its rover technology on to the Red Planet, aboard the Mars 2 and Mars 3 probes; however, both failed on landing.

NASA followed up Pathfinder with its twin Spirit and Opportunity rovers in 2004. Both kept going for years, but it was Opportunity that lived the longest, finally succumbing to a dust storm in June 2018, having trundled a total of 45.16 km (28.06 miles) across the Martian surface – to date, the furthest travelled by any robotic rover. They in turn were followed in 2012 by Curiosity, a car-sized rover designed to investigate the prospects for past life on Mars, as well as the planet's geology and climate.

NASA's Perseverance rover, which is based closely on Curiosity but with a different loadout of scientific instruments, is due to arrive shortly.

It's hoped this will set the stage for a Mars sample-return mission. Perseverance will actually make a start by gathering samples and leaving them in canisters dotted along its route, from where they could be collected by a future rover that will pack them into an ascent rocket for return to Earth. A sample returned to Earth could be subjected to the exhaustive battery of scientific tests possible in a full-size laboratory, rather than just the limited selection that can squeezed onto a spacecraft. Sample returns from the moon, the solar wind (the stream of particles billowing out from the sun), the tail of a comet and the surface of an asteroid have already been carried out by robotic spacecraft.

Outriders

The race to send robot explorers still further afield has also been well underway for decades. Even before the end of the Apollo moon programme, in March 1972 NASA launched Pioneer 10. It reached the asteroid belt beyond Mars later that year, and began returning photographs of the giant planet Jupiter and its moons in November 1973, before heading into the solar system's outer reaches. Contact was lost with the probe in 2003, and it's now estimated to be well in excess of 18 billion km (11 billion miles) from the sun. Pioneer 10 and its sister craft, Pioneer 11 (which also visited Jupiter, going on to become the first probe to fly past the ringed planet Saturn), both bear plaques showing human beings and their size compared to the spacecraft, and a diagram giving the location of Earth – just in case the probes should be intercepted by intelligent extraterrestrial lifeforms on their cruise through interstellar space after leaving the solar system.

> **Don't think of robots as replacements for humans – think of them as things that will help make us better at tackling many of the problems we face.**
>
> EOIN TREACY (2016)

Perhaps the jewel in the crown of outer-planet exploration was NASA's Voyager project, twin probes both launched in 1977. Voyager 1 encountered Jupiter and Saturn. But Voyager 2 was the star of the show, completing the Grand Tour, as it became known – flying past Jupiter, Saturn and Uranus, and culminating with a flyby of Neptune in 1989. Voyager's incredible imagery of the outer planets inspired a generation and its scientific findings set the agenda for the exploration of the outer solar system for decades to come.

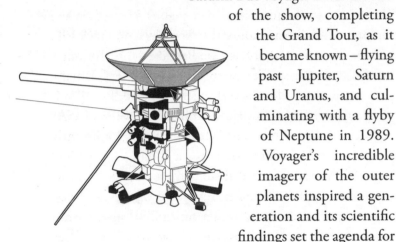

The Voyagers were followed in the 1990s and early 2000s by Galileo, which spent eight years exploring Jupiter and its moons; and Cassini, which did much the same for Saturn, and also deployed ESA's Huygens lander down to the surface of Saturn's largest moon, Titan. Huygens made the first ever landing on a moon of another planet, and the first landing anywhere in the outer solar system. The probe survived for 90 minutes on Titan's surface – where the average temperature is just –180 °C (–290 °F) – revealing a murky world of permanently frozen water ice and liquid hydrocarbons.

Pluto and beyond

Technically, Pluto ceased to be a planet in 2006, when the International Astronomical Union elected to reclassify it, and the thousands of icy bodies like it in the outer limits of the solar system, as 'dwarf planets'. Yet for many space enthusiasts, Pluto is still special. And so, fittingly, it finally got a visit from a spacecraft in 2015, when NASA's (obviously, robotic) New Horizons probe sped by at nearly 50,000 kph (31,000 mph). From a distance of 12,500 km (7,800 miles), it revealed surface features of Pluto, its largest moon, Charon, and its retinue of smaller satellites. New Horizons launched from Earth in 2006 and, thanks to a combination of low mass and a huge gravity assist (see Chapter 3) at Jupiter, quickly became the fastest spacecraft that's ever flown.

Along with the Pioneers and the Voyagers, New Horizons is now winging its way out of the solar system and towards the stars. In a sense, robots have explored here already. They've done this through the multitude of space astronomy probes that have been launched – craft that have journeyed into space to observe the distant universe unhindered by the obscuring haze of the Earth's atmosphere, or the light pollution and radio noise created by human activity.

The first was Explorer 7, a solar observatory sent into space by the United States back in 1959. Many more followed, to study the light from space in all its forms – from radio waves, through the visible spectrum, and right up to high-energy gamma rays. Probably the best known of these space observatories is the Hubble Space Telescope, which was launched into Earth orbit in 1990 and is still in operation today. Space observatories have shown us planets orbiting other stars, distant galaxies and supermassive black holes, and have even helped physicists to piece together the story of how our entire universe was born and evolved into its present state.

> If we are to send people, it must be for a very good reason – and with a realistic understanding that almost certainly we will lose lives.
>
> CARL SAGAN
> *Pale Blue Dot* (1994)

Meanwhile, some novel designs of space probe will be venturing to hitherto unvisited destinations. These include drone-like helicopters and quadcopters that will take to the skies of Mars and Saturn's mist-shrouded moon Titan. There are plans for submersibles to explore the liquid oceans believed to exist under the surface ice of some of Jupiter's moons. And 'Breakthrough Starshot', which we'll come back to later, is an ambitious plan to send an uncrewed probe on the four-light-year journey to the nearby star Alpha Centauri.

It seems that wherever humans travel in the solar system, and beyond, robots are, and will continue to be, our friends, colleagues and crewmates. Perhaps the science-fiction vision wasn't so wide of the mark after all.

06 SPACE IS BIG BUSINESS

'We're really at the dawn of a new era of space exploration, and one where there's a much bigger role for commercial companies.'

ELON MUSK (2012)

It's a beautiful sunrise – probably the best one today. You've seen five so far. A crescent of light gradually brightens until the sun bursts up into the blackness, sending a wave of colour tearing across the face of the planet 400 km (250 miles) below, picking out familiar landmasses and vast swirling weather systems.

You're watching from the comfort of a space hotel, an orbital tourist destination that circles the Earth once every 90 minutes. When you're not taking in the incredible sights, there's the opportunity for zero-G swimming, astronomy through the clearest sky you'll ever find and, for the truly intrepid, the chance to experience the outer limits first-hand with a spacewalk.

In 1967, Barron Hilton, then the co-chairman of Hilton Hotels, delivered a speech to the American Astronautical Society calling for the infrastructure to be created by which paying tourists could be ferried into space. Along with domestic robots and flying cars, it's become a dream long promised – but one that has so far failed to materialize. Now a number of space tourism companies are poised to make the dream very much a reality.

Some might argue that space hotels of a sort are already here. The first module of the International Space Station (ISS) was boosted into orbit in November 1998 aboard a Russian Proton rocket. In 2001 it became the first space station to accept paying guests, as the cash-strapped Russian space agency began flogging the seats allocated to Russian cosmonauts – much to the chagrin of sniffy NASA officials. Yet the costs are still way beyond the means of most.

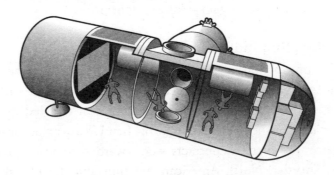

Space tourism is just one way in which the space environment is being exploited for commerce. Since the 1960s, Earth orbit has been filling with communication satellites, enabling corporations to beam data, phone calls and TV shows around the globe with ease. Satnav is another application that many of us wouldn't be without. But nowadays corporations aren't simply operating satellites – they're in the business of launching them, and sending their own space missions even further afield.

> **I am confident that when a space hotel becomes a practical reality it will also become a practical financial reality for Hilton.**
>
> BARRON HILTON (1967)

The perception of spaceflight being so difficult and expensive that only nation states dare attempt it has faded, giving way to a coterie of billionaires and private organizations who will fly people and payloads to the high frontier and back again – for the right price.

A fledgling industry

The production of spaceflight hardware has long been outsourced to the private sector. As far back as the early 1950s, and the design and construction of the Redstone ballistic missile that would take the first US astronauts into space, production contracts were awarded to firms such as Chrysler, North American Aviation and the Douglas

Aircraft Company. These companies simply manufactured and supplied vehicles and parts – they didn't own spacecraft, and certainly weren't permitted to fly them.

That began to change in 1962, when President John F. Kennedy signed the Communications Satellite Act into law. The act allowed private companies to own and operate Earth-orbiting satellites, effectively beginning the satellite communications industry. The first, Telstar 1, launched that same year and was a joint experimental project between AT&T, Bell Labs, NASA, the UK General Post Office and the French telecoms firm National PTT.

> **The ultimate purpose of space travel is to bring to humanity, not only scientific discoveries and an occasional spectacular show on television, but a real expansion of our spirit.**
>
> FREEMAN DYSON (1979)

Telstar departed the Earth from Cape Canaveral, aboard a NASA Thor-Delta rocket. The US space agency insisted on being involved and was actively opposed to letting private corporations meddle in space by conducting their own launches.

NASA didn't always get its way, though. In September 1982, a company called Space Services Inc. – whose staff included former NASA astronaut Deke Slayton – made the first ever privately operated rocket launch to space. Its Conestoga rocket was a modified Minuteman ICBM and it launched from Matagorda Island, situated 11 km (7 miles) off the Gulf coast of Texas, carrying a 500-kg (1,100-lb) test payload to an altitude of 309 km (192 miles). Unfortunately, it was the only successful launch that the rocket made.

NASA's attitude of resistance began to shift with President Ronald Reagan's Commercial Space Launch Act of 1984, which formally legalized privately operated space launches. And this was strengthened in 1990 with the Launch Services Purchase Act, which effectively forced NASA to procure commercial launch vehicles wherever possible.

Commercialization of space travel was also underway outside the United States. In 1980, the European Space Agency founded Arianespace, a company offering launch services to the space industry, initially just using its Ariane rockets, but now with a range of smaller boosters, allowing it to handle different-sized payloads. In the 1990s, Russia's spaceflight infrastructure was put on the market, with flights made available on its Soyuz and Proton rockets. After the retirement of the Space Shuttle in 2011, and before commercial crewed flight became available through US firms such as SpaceX and others, NASA had to book seats for its astronauts on Soyuz flights.

Girl from Mars

The prospects for private individuals to fly into space had been gathering pace through the 1980s, when NASA introduced the role of the 'payload specialist' on Space Shuttle flights. These crewmembers were often not career astronauts but scientists and engineers with technical expertise relevant to a particular mission – though they still had to undergo intensive astronaut training before they were permitted to fly.

In 1989, a consortium of UK companies – including British Aerospace and Interflora – launched Project Juno, to put a British citizen in space. There were 13,000 applicants for the job ('Astronaut wanted. No experience necessary') of which four were selected for eighteen months of training at Star City, Russia's cosmonaut boot camp. Of the four, twenty-six-year-old Helen Sharman – a chemist who had been working as a food technologist for confectionery firm Mars – was eventually selected to fly. On 18 May 1991, she boarded a Soyuz rocket and took off for the Russian Mir space station, where she spent eight days carrying out scientific experiments, photographing Britain from space, and taking time to speak via radio with children back on Earth. Upon her return, Sharman became a popular science communicator, writing and lecturing about her experience. In 1990, Toyohiro Akiyama, a Japanese journalist, also flew to Mir on a Soyuz spacecraft, spending a week on the station, on a trip funded by his employer, the Tokyo Broadcasting System.

The 1990s was a financially stressful time for the Russian space programme, owing to the state of the country's economy after the collapse of the Soviet Union in 1991. Space tourism was seen as one way to raise some much-needed funds, which led the Russian space agency to partner with private investors to create MirCorp, a company whose aim was to generate revenue from Mir. One strand of its business plan was tourism.

MirCorp's first booking was the American businessman Dennis Tito, who announced in June 2000 that he had paid $20 million for a stay on the space station. However, he was too late and Mir was abandoned in 1999, and then de-orbited in March 2001, burning up in the Earth's atmosphere over the Pacific Ocean. Tito still got his trip though, flying instead to the International Space Station at the end of April 2001, where he spent eight days in orbit. Although Sharman and Akiyama preceded him, Tito is generally regarded as the first space 'tourist', since he was the first to pay for his own ticket.

> The true future of space travel does not lie with government agencies ... but real progress will come from private companies competing to provide the ultimate adventure ride.
>
> BUZZ ALDRIN (2009)

Eye on the prize

Six more visitors to the ISS followed and one of them, the Hungarian-American software engineer Charles Simonyi, even flew twice. Yet the price has not diminished greatly. That looks set to change very soon though, with Virgin Galactic about to start taking passengers on short, sub-orbital hops above the Kármán line.

Their vehicle, called SpaceShipTwo, flies up to an altitude of 15,000 m (50,000 ft) slung beneath a conventionally powered jet aircraft. From here, it's dropped and ignites its hybrid rocket engine, rapidly accelerating the craft to 4,200 km/h (2,600 mph, or 3.4 times sound speed), sufficient for it to coast up to a peak altitude of 110 km (68 miles). During the coasting phase, the craft is essentially in freefall, allowing passengers to experience around six minutes of weightlessness.

No heat shield is required for re-entry to Earth's atmosphere on a sub-orbital flight – instead, the Virgin 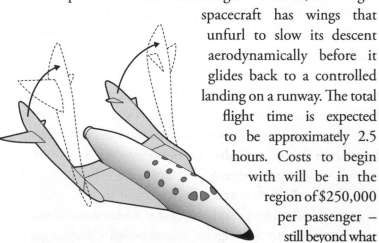 spacecraft has wings that unfurl to slow its descent aerodynamically before it glides back to a controlled landing on a runway. The total flight time is expected to be approximately 2.5 hours. Costs to begin with will be in the region of $250,000 per passenger – still beyond what most of us can realistically afford, though the fare is expected to drop once the technology matures.

Virgin's flights depart from Spaceport America, an FAA-licensed launch and landing facility in the US state of New Mexico, though other spaceports exist in Oklahoma, California, Alaska and elsewhere. Virgin had anticipated starting to launch paying passengers as early as 2007. But several incidents have pushed the date back. In October 2014, a Virgin Galactic spacecraft broke up in mid-air during a test flight, killing one of its pilots. The incident is thought to have been caused by the spacecraft's wings accidentally being deployed for atmospheric re-entry while its main rocket was still firing. During re-entry, the wings create a large aerodynamic drag force, used to slow the spacecraft's descent – but during powered flight this force would become too great for the vehicle's structure to withstand.

Now the programme seems to be back on track. In February 2019, a Virgin spacecraft made its first flight with three astronauts on board – two pilots plus a passenger (the full complement is eight – six passengers and two pilots).

Virgin Galactic's technology is licensed from California-based aerospace firm Scaled Composites, who developed the spacecraft's forerunner, SpaceShipOne, to win the $10 million Ansari X Prize – a contest to privately finance, build and launch twice in a period of just two weeks, a spacecraft carrying people to the edge of space. Scaled Composites made its second flight, and claimed the prize, on 4 October 2004 – the forty-seventh anniversary of Sputnik 1.

The prize was co-funded by the Iranian-American entrepreneurs Anousheh and Amir Ansari, and the X Prize Foundation – a not-for-profit organization that fosters technological innovation through healthy competition. The foundation was inspired by the Orteig Prize, a $25,000 reward put up in the early twentieth century by hotelier Raymond Orteig, for the first non-stop flight across the Atlantic Ocean. It was claimed in 1927 by US aviator Charles Lindbergh, who made the 5,800 km (3,600 mile) crossing from New York to Paris in 33.5 hours in his Spirit of St Louis monoplane.

Fly me to the moon

While Virgin Galactic's activities are confined to the narrow region just above the Kármán line, other private spaceflight organizations have their sights set further afield. Corporations such as SpaceX, Blue Origin and the United Launch Alliance (a joint space launch venture by aerospace manufacturers Lockheed-Martin and Boeing) are building and flying spacecraft that can reach orbit, the moon, and possibly Mars and beyond.

SpaceX in particular seems to have made astonishing progress in a relatively short time. The company was created in 2002 by technologist and businessman Elon Musk, with his proceeds from the sale of online payment provider PayPal, which he had co-founded. In 2008, SpaceX's liquid-fuelled Falcon 1 rocket reached orbit. In 2012, the more powerful Falcon 9 sent a capsule to dock with the International Space Station – a first for a private corporation. It is also the first company to autopilot a rocket stage back to a vertical landing on Earth so that it can be re-used – reducing the cost of each launch to $50m (similar-sized expendable rockets cost twice this).

And in 2018 SpaceX made the debut launch of its Falcon Heavy booster, which as a publicity stunt carried Musk's car – a red Tesla Roadster – and threw it onto an orbit around the sun, another first for a private company.

SpaceX is now using the Falcon 9 to fly uncrewed re-supply missions to the ISS and, at the time of writing, has just validated its Dragon capsule for human spaceflight, so that it can ferry astronauts to and from the station. The first launch of Crew Dragon was scheduled for May 2020, and so may well have taken place already by the time you read this. NASA has also stated that it will allow private astronauts on the ISS, flying there aboard Crew Dragon, for a cost of US$35,000 per day.

> **In order to reach the space station, we will work with a growing array of private companies competing to make getting to space easier and more affordable.**
>
> PRESIDENT BARACK OBAMA (2010)

TIMELINE OF COMMERCIAL SPACEFLIGHT

Date	Description
1962	Telstar 1, the first commercially operated satellite – run by AT&T, Bell labs, and others – is launched from Cape Canaveral
1982	Conestoga 1 is the first privately owned and operated rocket to make it into space, delivering its payload to an altitude of 309 km (192 miles)
1990	The air-launched Pegasus rocket, operated by Orbital Sciences Corporation, delivers a payload to low Earth orbit after being dropped from an aircraft like a cruise missile
2000	Russian Soyuz TM-30 flight to the Mir space station is the first privately funded mission to an Earth-orbiting space station
2001	Dennis Tito becomes the first paying space tourist when he shells out $20 million for a stay aboard the International Space Station
2004	SpaceShipOne makes the first private crewed flight into space. Later that year it makes two such flights within a week, winning the Ansari X Prize
2008	The SpaceX Falcon 1 becomes the first privately funded liquid-fuelled (and therefore able to be throttled, making it suitable for human travellers) rocket to reach orbit
2015	President Barack Obama makes it legal for private companies to own any resources that they may recover from space

Date	Description
2018	SpaceX makes the maiden flight of its Falcon Heavy rocket, made of three Falcon 9s strapped together. It's the third most powerful rocket ever to fly, behind the US Saturn V and Russian Energia
2019	SpaceX's Starhopper, a test vehicle for its interplanetary transport technologies, makes a controlled flight to 150 m (492 ft) and then lands safely

SpaceX's next major spaceflight project is called Starship. And it's straight out of science fiction – a 50-metre-long (165 feet) stainless-steel rocket ship capable of carrying up to a hundred people, or the equivalent in cargo, to Earth orbit and on to other planets of the solar system. Another variant will serve as a tanker, carrying propellants to Earth orbit to provide in-flight refuelling. All the Starship variants will be launched from Earth atop a rocket called the Super Heavy – an utter behemoth, expected to deliver twice as much thrust on take-off as the Saturn V rockets that powered the Apollo missions.

On other worlds, such as the moon and Mars, the surface gravity is much lower, allowing Starship to take off without the assistance of the Super Heavy. In August 2019, a test vehicle called Starhopper, resembling a large metal water tank on legs, flew to a height of 150 metres (490 feet), hovered and made a controlled landing. As of early 2020, two full Starship prototypes were under construction. Once operational, Starship will replace the Falcon rockets as the company's default mode of reaching space, though this isn't expected to happen until 2021 – and even that date still seems ambitious, given the immense scale of the project. Musk argues that the cost per launch could eventually drop to as low as $2 million, a tiny fraction of current prices.

There has already been interest in Starship from potential tourists. Japanese billionaire Yusaku Maezawa wants to charter a private flight in 2023 to swing around the far side of the moon and back – a so-called free-return trajectory – taking a number of artists with him to capture the experience. Called dearMoon, the project could be the first crewed space flight beyond low Earth orbit since 1972. Musk has stated that he intends Starship to fly to Mars, perhaps as soon as 2025 (though, again, this seems optimistic), with the ultimate aim of establishing colonies there.

Spaceflight Inc.
Even with a refuelling stop in Earth orbit, a Mars mission would still require additional fuel for the return journey,

which it's hoped could be manufactured on Mars's surface, through a technique called 'in situ resource production'. This would involve drilling up sub-surface water ice (H_2O) on Mars and combining it with carbon dioxide (CO_2) in the planet's atmosphere to produce the liquid oxygen (O_2) and liquid methane (CH_4) that Starship burns as fuel.

Others have proposed mining chemicals and minerals from other worlds and returning them to Earth. While the 1967 Outer Space Treaty prevents any nation state from claiming sovereignty over a celestial body, there are no restrictions governing the recovery and exploitation of resources. In November 2015, President Obama signed the Commercial Space Launch Competitiveness Act – allowing private US firms and citizens to 'stake their claim' to, and profit from, any inanimate materials that they are able to retrieve from space.

Targets could include precious metals – such as gold, silver and platinum – and elements that are rare on Earth but plentiful in space, such as iridium and rhenium, which could potentially be mined from asteroids. Rocket fuel is another asset that, as on Mars, can be manufactured from water ice on the moon and then boosted into space for refuelling.

> **Are governments the only entities that can build human spacecraft? No – actually, every human spacecraft ever built for NASA was built by private industry.**
>
> ALAN STERN (2010)

Because of the low lunar gravity, it's much cheaper to send it from the moon than from Earth.

A prime candidate for lunar mining is helium-3, a type of helium in which each atom has a neutron removed from its nucleus (left in the diagram below). Deposits of this substance are believed to have been trapped in lunar soil from the solar wind – the stream of particles steadily billowing out from the sun. Helium-3 could serve as a fuel for future nuclear fusion reactors, which generate energy by sticking atoms together – much like the nuclear reactions that take place inside the sun. Many potential fusion fuels give off deadly and hard-to-contain neutron radiation, but with helium-3 this is not the case, making it a highly desirable commodity to retrieve and bring back to Earth.

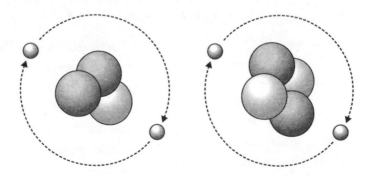

The lesson from the history books is that wherever the pioneers go, industry seems to follow. That's what we're seeing as the utilization of the near-Earth environment gets underway, and commerce becomes its own driver for the expansion of humanity into space. Yet space hotels, asteroid mining and privately funded excursions to other planets are, for the moment, all fantastic adventures for the future. There are, though, some incredible missions, bound for some incredible destinations, that are happening right now. And that's where we're off to next.

07 THE NEXT GIANT LEAP HAS ALREADY BEGUN

'We will accelerate our return to the moon by 2024 and establish a foundation for a sustainable human presence by 2028.'

JIM BRIDENSTINE, NASA ADMINISTRATOR (2019)

Now is a heady time to be a space enthusiast. NASA's Perseverance rover is the size of a car and incorporates a helicopter scout drone to conduct aerial surveys. It is due to land on Mars in early 2021. Its primary mission is to investigate the possibility that Mars was once a living world. Also happening over the coming decade are probes to return samples from asteroids and to search for oceans on Jupiter's icy moons, while the much-anticipated James Webb Space Telescope, the successor to the mighty Hubble, will finally take to the skies.

But the robots are set to get a run for their money, as there are also going to be some exciting developments in crewed space exploration over the coming years.

2020 vision

The Red Planet, Mars, will continue to be a focal point for space exploration this decade. Although not the closest planet to Earth, in terms of either size or proximity (that honour goes to Venus on both counts), Mars is the most Earth-like of all the other worlds in the solar system. It has a rocky surface, surrounded by a transparent atmosphere. It has appreciable gravity, more than on the moon – about one-third of Earth's. And it has weather systems, volcanoes and water. Looking at a picture of Mars's surface, it's not impossible to imagine it might have been taken in an arid desert somewhere on Earth. Mars is a beguiling world that has long teased us with its secrets, not least the possibility that life may once have existed there. And that's why NASA is sending its most ambitious robotic lander mission yet.

Called Perseverance, it's a gigantic wheeled rover, weighing over a tonne (1.1 US tons) and measuring almost 3 metres (3.3 feet) in length (about the size of a

> **We're going to launch a series of missions to begin that search for life on Mars and now is more appealing than ever before.**
>
> DR JIM GREEN
> NASA chief scientist (2015)

small car). The mission will blast off from Cape Canaveral aboard a powerful Atlas V rocket, with a launch window of 17 July to 5 August 2020. The transfer orbit to Mars takes about seven months to traverse, with touchdown scheduled for 18 February 2021. The chosen landing site is Jezero Crater, believed to have been a lake bed between 3.5 and 4 billion years ago. The sediments deposited in the water that once flowed here may harbour the chemical signatures of any life that was present on Mars at the time, as well as actual preserved fossils.

Among its many scientific instruments, the rover is packing a drill that can extract cores from the Martian surface. These rock cores will be left on the planet's surface in capsules, for collection by a future rover, which will transfer them to an ascent vehicle for return to Earth (though this mission has yet to be approved). Perseverance will also road-test technology for extracting oxygen from the Martian atmosphere, necessary for the establishment of any long-term human presence on the planet.

Coming into Mars's atmosphere at 26,000 km/h (16,000 mph), a heat shield will protect the lander as it brakes aerodynamically, heating its exterior to 2,100 °C (3,800 °F). The atmosphere on Mars is dense enough that the heat shield is required, but equally it's too thin for the aerobraking alone to slow the craft down fully. Still travelling at a speed of around 1,400 km/h (900 mph), parachutes will then deploy to slow its descent further, to 300 km/h (200 mph), before a cluster of retro rockets fires. Mars is too far from Earth for ground controllers to remote-pilot the lander down. Instead, software, guided by radar sensors, will autopilot the descent vehicle to the surface, where the rocket engines bring it to a near-hover at an altitude of about 20 metres. From here, a winch lowers the rover down to the surface, pyrotechnics cut the cables, and the descent vehicle separates and flies clear to a crash landing. Making a powered descent all the way to the planet's surface isn't an option, as this would throw up clouds of dust, which could damage the rover's delicate instruments.

This seemingly rather Heath Robinson landing method is called a 'sky crane' and it worked perfectly for the touchdown of Perseverance's forerunner, the Curiosity rover, in 2012. Airbags, the landing system used on previous rover missions, simply aren't strong enough to withstand the impact of a spacecraft this large and heavy. The whole entry, descent and landing phase of the mission takes approximately seven minutes and is extremely complex and precarious. So much so that NASA engineers refer to it as the 'seven minutes of terror'.

Rock and roll

Once down on the ground, the rover will begin its mission, with an initial planned duration of one Martian year (1.88 Earth years). This was also the conjectured primary mission duration for Curiosity, which, at the time of writing, is into its fifth Martian year and still going strong. During this period, Perseverance is expected to drive roughly 200 metres (650 feet) per day, examining the soil and rocks that it encounters. And with no fewer than twenty-three cameras on board, you can guarantee there will be no shortage of stunning imagery to accompany the science results.

After sixty days, Perseverance will deploy a small helicopter drone to begin a short programme of test flights. It won't do any actual science – its mission is purely to investigate flying on other worlds – but it will carry two cameras. Aerial images taken by the helicopter will be many times the resolution of those returned from orbit. Future aircraft explorers could provide surveys to guide rovers on the ground and fly to inaccessible locations, such as

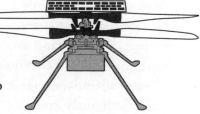

THE LUNAR GATEWAY

In 2022, NASA – in partnership with ESA, Roscosmos and the Canadian and Japanese space agencies – plans to launch the first component of the Lunar Gateway, a crewed space station in orbit around the moon. The space station will adopt a multi-module design, with a pressurized internal volume of 125 cubic metres (4,400 cubic feet) compared to 915 cubic metres (32,333 cubic feet) on the ISS. It will include modules for power and propulsion, habitation for the four-person crew, as well as scientific equipment, and docking facilities for visiting spacecraft. The station will be placed into a six-day elliptical orbit passing over the moon's poles, taking it out to a greatest distance of 70,000 km (43,000 miles) with a closest approach of 3,000 km (1,900 miles). The Lunar Gateway is a key part of the Artemis programme, a project to take astronauts back to the lunar surface. It will provide a reusable base from which to conduct lunar exploration and research, and will serve as a jumping-off point for sending crewed missions further afield.

cliffs. It will be the first time a powered aircraft has been deployed on another world.

The helicopter is powered by a lithium-ion battery, which can be recharged in between flights by onboard solar panels. The rover itself is supplied with electricity from a radioisotope thermoelectric generator (or, RTG for short – see Chapter 8), containing 4.8 kg (11 lb) of plutonium dioxide. This nuclear battery has a lifetime of fourteen years. Solar panels are unable to meet the power demands of such a large vehicle and its instruments, and are prone to getting covered with Martian dust, which reduces their efficiency.

Ex machina

We'll also be visiting the moons of Mars over the coming years. In September 2024, the Japanese space agency, JAXA, plans to launch the Martian Moons Exploration (MMX) robotic probe to study Mars's two natural satellites, Phobos and Deimos. The spacecraft will actually land on Phobos, collecting 10 g (0.35 oz) of soil samples, and then take off again, making flybys of Deimos before launching the sample capsule back to Earth, with a planned arrival date of July 2029.

Sample-return missions are in vogue at the moment. As well as the planned mission to collect and return the cached samples gathered by the Perseverance rover, there are other projects already in flight. NASA's OSIRIS-REx is, at the time of writing, in orbit around the near-Earth asteroid 101955 Bennu. Before leaving, it will make a very close flyby, extending its robot arm down to collect a surface sample, which it will bring back to Earth in September 2023. Another asteroid sample-return probe, JAXA's Hayabusa2, is already en route home with material from asteroid 162173 Ryugu, both surface soil and rock excavated from within the asteroid. It's due to return to Earth in late 2020. Meanwhile, China is readying its Chang'e 5 and 6 probes. Beginning in December 2020, they will head to the moon to return samples in preparation for establishing a permanent research outpost near the lunar south pole.

> **JUICE will give us better insight into how gas giants and their orbiting worlds form, and their potential for hosting life.**
>
> ALVARO GIMÉNEZ CAÑETE
> ESA (2012)

The search for life elsewhere in the solar system continues, and it's not just Mars that has caught the attention of scientists. Three of Jupiter's large moons – Callisto, Europa and Ganymede – have icy outer surfaces that are thought to conceal oceans of liquid water beneath, heated and kept liquid by the repeated squashing and squeezing effect of Jupiter's gravitational field. And this has naturally led to speculation over the possibility of active extraterrestrial biology there. To investigate, ESA launches its Jupiter Icy moons Explorer (JUICE) mission in June

2022. Arriving in October 2029, the probe will embark upon a tour of the Jovian system, making flybys of Europa and Callisto, and finally entering orbit around Ganymede. It will use an ice-penetrating radar to investigate the presence of sub-surface water, and determine the thickness of the covering ice. The mission will likely coincide with NASA's Europa Clipper orbiter. Between them, the two spacecraft will lay the groundwork for a future robotic lander mission to these faraway sub-worlds.

Far and wide

NASA has plans to head even further afield, launching a lander mission called Dragonfly to the icy world Titan, the largest moon of the ringed gas-giant planet Saturn. Titan was the first ever moon of another planet to be visited by a lander craft when ESA's Huygens probe set down on 14 January 2005. Huygens was a static lander, and survived for just 1.5 hours on Titan's −180 °C (−290 °F) surface. In contrast, Dragonfly will be a drone-like quadcopter, which, powered by a nuclear RTG unit, will be able

> It's remarkable to think of this rotorcraft flying miles and miles across the organic sand dunes of Saturn's largest moon, exploring the processes that shape this extraordinary environment.
>
> THOMAS ZURBUCHEN
> NASA (2019)

THE JAMES WEBB SPACE TELESCOPE

Launching in March 2021, the James Webb Space Telescope (JWST) is a large space-based infrared telescope that will shed light on the formation of stars, planets and even galaxies. Infrared astronomy is a powerful technique, because infrared light – which has a longer wavelength than visible light – can pass through obscuring clouds of gas and dust to reveal details about the universe that would otherwise be hidden from view. It's absorbed by the Earth's atmosphere, though, which is why infrared telescopes must be situated in space. The JWST will peer into cosmic gas clouds where stars and planets are believed to be in the process of forming. In cosmology, light is redshifted (stretched out to longer wavelengths) by the expansion of the universe, making infrared the best way to observe the most distant cosmic objects. JWST will see so far out into space (and, therefore, back in time thanks to the delay created by the finite speed at which light travels) that it will capture images of the very first generation of galaxies to form in the universe.

to soar to an altitude of up to 4 km. It will hop from site to site across the moon, investigating its lavish carbon-based chemistry, which is believed to mirror that of the very early Earth, and as such may offer clues as to how life first emerged on our planet. Dragonfly is expected to launch in 2026 and arrive at the Saturn system in 2034, touching down in Titan's Shangri-La dune fields.

Not everything in space that's studied by a robotic spacecraft gets an up-close visit. One of the most hotly anticipated upcoming missions is the James Webb Space Telescope (JWST), the successor to the Hubble Space Telescope. Named after the NASA administrator who oversaw the agency's Apollo era during the 1960s, the JWST will sit at the L2 Lagrange point (see Chapter 3) on the far side of the Earth from the sun. Here, shaded from the solar glare, it will scrutinize the distant universe. The telescope's main mirror measures 6.5 metres (21 feet) across – nearly three times the size of Hubble's, so big that the mirror is broken into segments that fold away for launch, and then unfurl in space.

The Earth-orbiting Hubble, having been deployed with a fatal flaw in its optics, was visited by a crewed repair mission. However, at 1.5 million km (930,000 miles) from Earth, there will be no such option with the JWST, so the pressure is on to get it right first time. The mission will launch from Earth aboard an ESA Ariane 5 rocket on 30 March 2021, and take up its

> **The James Webb Space Telescope was specifically designed to see the first stars and galaxies that were formed in the universe.**
>
> JOHN GRUNSFELD (2017)

station at the L2 Lagrange point some thirty days later. Its initial science mission is planned for five years, with the possibility to extend this to ten – the principal limitation being the fuel required to maintain the telescope's position at L2.

Crew's missile

NASA is planning a dramatic return of human astronauts to the moon. The project, called the Artemis programme (Artemis was the twin sister of Apollo in Greek mythology), is due to make its first uncrewed test flight in 2021. The spacecraft is called Orion, and will be launched atop the somewhat cumbersomely titled Space Launch System (SLS) super-heavy-lift rocket. The flight will put the spacecraft through its paces, lasting over twenty-five days and including six days of orbits around the moon.

THE NEXT DECADE IN SPACE

Date	Mission	Description
2020	Hayabusa2	With samples of asteroid 162173 Ryugu safely on board, the Japanese Hayabusa2 mission is due to return to Earth in late 2020
18 Feb 2021	Perseverance	NASA's next Mars rover mission will touch down in Jezero Crater in the northern hemisphere, to investigate the planet's geology and past habitability
30 Mar 2021	James Webb Space Telescope	The space telescope to replace Hubble packs a 6.5-metre mirror capable of seeing the universe as it was just 200 million years after the Big Bang
Dec 2021	Gaganyaan	The inaugural human spaceflight of the Indian Space Research Organization (ISRO) will be a seven-day Earth-orbiting mission
2022	Tiangong-3	Construction will begin on China's third crewed space station, an Earth-orbiting platform built from multiple modules, like the ISS and Mir, in 2022
2022–28	Artemis programme	NASA's much-vaunted return of human astronauts to the moon will see a total of seven crewed missions visit the lunar surface
2026	Dragonfly	After arriving in 2034, this quadcopter drone will spend three years exploring Saturn's largest moon, Titan, from the sky
2028	Martian Moons Exploration	MMX, operated by the Japanese space agency, will launch in 2024 for the Martian moon Phobos and collect samples for return to Earth in 2028
2029	Jupiter Icy moons Explorer	Launching in 2022, ESA's JUICE mission will travel to Jupiter to study its frozen moons, which are thought to have liquid water beneath their surfaces

If successful, it will be followed in September 2022 by Artemis 2, carrying a crew of four on a free-return (figure-eight) trajectory to loop around the back of the moon and then return to Earth over the course of a ten-day flight. Artemis 3 will be the mission we've all been waiting for. Currently slated to depart Earth in 2024, it will attempt the first crewed landing on the moon in over fifty years. NASA will be partnered in the Artemis programme by JAXA, ESA and the Canadian Space Agency. A further four Artemis missions are set to follow in the timeframe 2025–28. Each a month in duration, these will continue human science activities on the moon's surface and begin construction of the Lunar Gateway space station. NASA currently aims to land its own human crews on Mars by 2033, and the experience gained from Artemis will be crucial.

Artemis will be sparring with SpaceX and others in the human conquest of outer space. But other nations are dipping their toes in the cosmic ocean too. India plans to launch its first crewed space mission in 2021. Its Gaganyaan spacecraft will carry a crew into Earth orbit, and sustain them there for a week. And China, already

> **By December 2021, the first Indian will be carried by our own rocket … This is the target ISRO is working for.**
>
> KAILASAVADIVOO SIVAN
> Chairman ISRO (2019)

a spacefaring nation, is expected to launch Tiangong-3, its third crewed space station, by 2022. This achievement will surely cement China as the world's third major space-launching power, after the United States and Russia.

This coming decade promises to be the most exciting time for spaceflight since the Apollo programme of the 1960s. We truly are the lucky ones to have a ringside seat for what promises to be the greatest show on Earth – and way, way beyond.

08 IT'S A SMALL SOLAR SYSTEM

'The time was fast approaching when Earth, like all mothers, must say farewell to her children.'

ARTHUR C. CLARKE
2001: A Space Odyssey (1968)

Crossing the yawning gulf of interplanetary space that separates the Earth from the other worlds of the solar system is an arduous trek to make. At present, a trip from Earth to the ringed gas giant Saturn takes eight years – and getting to the outer reaches of the solar system even longer.

Moving through outer space this slowly isn't just inconvenient – it's dangerous. Space is an extremely perilous place for human explorers, with hazards ranging from deadly radiation, to the airless cold vacuum, to our dependence on myriad life-support technologies, all of which have to continue working perfectly to keep a human being alive. Put simply, the longer you stay in space the more likely it is that something will go catastrophically, life-threateningly wrong.

The problem is our current rocket technology. Rockets work by combustion, a chemical reaction where flammable fuel is combined with oxygen to burn, releasing energy which then accelerates the fuel through the engine's nozzle at very high speed, pushing the rocket in the opposite direction. But chemical combustion is a relatively low-energy process, meaning that rocket engines must guzzle a huge amount of fuel, all of which itself must be carried into space with the rocket, making them rather sluggish.

Luckily for us, chemical rockets are far from the last word when it comes to space-propulsion systems. In the decades to come, we can expect to see craft propelled by nuclear engines, while a new kind of ion drive could cut the trip to Mars from months to just a few weeks.

And there are other ideas too, such as 'solar sails' – giant reflective sheets resembling tin foil, enabling a spacecraft to hitch a ride on the light streaming out from our star. Even the way in which we leave Earth and access space is set to change. Scientist and author Arthur C. Clarke, for example, was a strong advocate of 'space elevators', cables suspended from orbiting platforms to winch payloads up from the Earth's surface and into space.

ARTHUR C. CLARKE (1917–2008)

Arthur Charles Clarke was born in Minehead, Somerset, on 16 December 1917. During the Second World War, he worked on the development of radar for the Royal Air Force. In 1945, a paper he published in *Wireless World* magazine set out the principles needed to use satellites for global communications. Geostationary satellite orbits are still sometimes referred to as 'Clarke orbits' today. Following the war, he obtained a degree in maths and physics from King's College London. Soon afterwards, Clarke became president of the British Interplanetary Society and began publishing popular books on space travel. He had begun writing science-fiction stories in the 1930s and published his first novel, *Against the Fall of Night*, in 1948. Clarke focused on 'hard' science fiction – stories based on solid, or at least plausible scientific concepts. He is best known for *2001: A Space Odyssey*, published in 1968, while his 1979 novel *The Fountains of Paradise* popularized the concept of the 'space elevator' (see page 133). Clarke was knighted in 2000 for services to literature. He died at his home in Colombo, Sri Lanka, on 19 March 2008.

Nuclear age

The use of nuclear technology to power spacecraft dates right back to the early days of the space programme. In fact, as far back as the Second World War, scientists on the US Manhattan Project, seeking to build the first atomic bomb, had begun to speculate on the application of the technology to rockets. A project was started in the early 1950s at Los Alamos National Laboratory, New Mexico. Called Project Rover, its brief was to develop a nuclear-powered upper-stage engine for the US Atlas intercontinental ballistic missile. However, by the late fifties it was clear that this would be overkill for Atlas. But at the same time, scientists were beginning to appreciate the relevance of nuclear power for the emerging space programme. In August 1960, NASA created the Space Nuclear Propulsion Office (SNPO). Its principal job was to oversee Project Rover, now renamed NERVA (Nuclear Engine for Rocket Vehicle Application).

NERVA was a 'nuclear thermal rocket', meaning that energy from nuclear reactions (rather than chemical combustion) is used to directly heat the rocket's fuel. The basic idea is to pump hydrogen through a hot nuclear reactor, heating it to over 2,200 °C (3,990 °F). At this temperature, the gas can then be passed through a rocket nozzle to form a high-speed exhaust moving at 7 km/s (4.3 miles per second) and propel the spacecraft.

The reactor itself operated on the principle of nuclear fission. This is the same technology used in nuclear power stations, and works by splitting large, heavy atoms – in this case, uranium – to release energy. NERVA's reactor generated an output of 1,137 megawatts – enough power to boil over 600,000 electric kettles simultaneously.

The first firing of a full, working NERVA engine took place on 24 September 1964, at the project's Jackass Flats test site, in Nevada. Ground tests continued successfully throughout the 1960s. There was even a case made at one point for NERVA to provide upper-stage propulsion for the Apollo missions, but it was thought unlikely the technology would be operational in time to hit President Kennedy's target of reaching the moon by the end of the decade.

Instead, NASA opted to keep NERVA for a potential Mars mission in the late 1970s, and it was also a candidate to power the Voyager missions on their grand tour of the outer solar system. But, alas, none of it was to be. With the huge expense of Apollo and the spiralling cost of the Vietnam war, President Nixon cancelled the project in January 1973. And to date, NERVA has yet to fly.

> **As we push out into the solar system, nuclear propulsion may offer the only truly viable technology option to extend human reach to the surface of Mars and to worlds beyond.**
>
> SONNY MITCHELL
> NASA (2017)

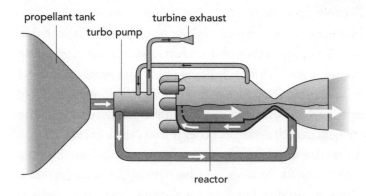

propellant tank | turbine exhaust | turbo pump | reactor

Now, however, nuclear thermal propulsion could be making a comeback. At the National Space Council meeting in Chantilly, Virginia, on 20 August 2019, NASA Administrator Jim Bridenstine called for a revival of the old NERVA technology. And in May the same year, the US House Appropriations Committee approved US$125 million for NASA to conduct research into nuclear thermal propulsion systems. A team at the Marshall Space Flight Center, in Huntsville, Alabama, is overseeing the research, and contracts to produce hardware have already been awarded to industry.

NASA is now developing the safety protocols necessary for the launch of a nuclear reactor into space. With the Artemis programme set to take astronauts back to the moon, and the Lunar Gateway serving as an in-flight waystation to deep space, it's thought NERVA-type engines may get a chance to prove their worth as soon as 2024.

Power up

Nuclear thermal engines are pretty much just rockets that heat their propellant with a nuclear reactor rather than setting fire to it. Another possibility is to use the energy produced by the reactor to drive an electric propulsion system. We've seen how ion drives can do this, by electrically accelerating charged particles – ions, as they're also called. Space missions fitted with ion drives have been flown in space since the late 1990s, and there have now been more than half a dozen of them, making ion drives a thoroughly tried and tested propulsion technology. All of these have been solar-powered, converting light from the sun into electricity, which in turn drives the engine. But this means that these craft must confine their activities to the vicinity of our star.

Probes to the outer solar system, far from the sun's light and heat, already take their own power with them. The Galileo and Cassini robotic missions to Jupiter and Saturn, while propelled by conventional chemical rockets, carried radioisotope thermoelectric generators (RTGs) to power

> **For decades, people have known the chemical-propulsion approach to space travel is really not going to get us that far. Chemical propulsion is essentially like the horse-and-cart approach to the exploration of the American West, instead of the steamboat or the railroad.**
>
> FRANKLIN CHANG DIAZ (2009)

their communications systems and scientific instruments. RTGs are essentially small lumps of radioactive material – such as plutonium oxide. The radioactivity causes atoms within the material to break apart spontaneously and create heat, which is then converted into electricity. But RTGs are quite primitive devices and there's a limit to how much power can be wrung from them. Carrying a full-spec nuclear reactor on board is a far more satisfactory solution, giving a spacecraft enough juice to operate all of its basic instruments and, if needed, an electric propulsion system too – and do so almost anywhere.

In years to come, nuclear electric propulsion could really have its day thanks to a new kind of electrically powered thruster called VASIMR, which stands for Variable Specific Impulse Magnetoplasma Rocket. 'Specific impulse', as we saw back in Chapter 2, is the amount of thrust a rocket can deliver per unit mass of fuel burned – and so the higher it is, the more efficient the engine. For technical reasons, specific impulse is measured in seconds. A Space Shuttle main engine (which is a chemical rocket burning liquid oxygen and liquid hydrogen) has a specific impulse of about 450 seconds, which it achieves by accelerating its exhaust gases to in excess of 4,000 m/s (13,000 ft per second). An ion engine can improve on this significantly, cranking the specific impulse up to around 3,000 seconds, with an exhaust speed of up to 50,000 m/s (160,000 ft per second). But VASIMR has the potential to take this off the scale. Its specific impulse can be as high as 12,000 seconds, which it achieves from a staggeringly fast exhaust velocity of 120,000 m/s (395,000 ft per second).

A conventional ion engine makes ions by knocking electrons from atoms, leaving the atoms electrically charged so that they can be accelerated by an electric field. VASIMR is different, using radio waves to convert a cold gas into a plasma – a superheated gas of ions – at a temperature of more than 1 million °C (1.8 million °F). Powerful electromagnets, made from superconducting wire (with zero electrical resistance) to maximize their efficiency, then confine and funnel the plasma into an exhaust jet – effectively acting as a kind of 'magnetic rocket nozzle'.

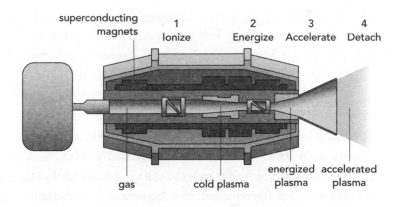

VASIMR was originally developed to create high-temperature plasmas for nuclear fusion – a method for generating nuclear power that works by welding together lighter atoms to release energy. Fusion is the power source that operates in the searing hot core of the sun and other stars, but it requires high temperatures to kick-start it, and that's where VASIMR comes in. The technology is now being adapted for spaceflight by the Texas-based Ad Astra Rocket Company, run by former NASA astronaut Franklin Chang Diaz. In 2008, the company successfully ground-tested a 200-kilowatt VASIMR engine, and in 2015, it received a three-year US$9m grant from NASA to develop the concept for a possible flight test.

> **You can do chemical propulsion to Mars, but it's really hard. Going further than the moon is much better with nuclear propulsion.**
>
> BILL EMRICH
> NASA (2019)

Like ion drives, VASIMR is an 'in space' propulsion technology, to be used once a spacecraft has broken free of the Earth's gravity. But whereas ion engines have fixed specific impulse, with VASIMR it's adjustable – allowing the engine effectively to 'change gear', trading off thrust against efficiency and back again as conditions require. In deep space, a high-efficiency, low-thrust configuration works best, gathering speed gradually, accumulating 'delta-v' by firing continuously for long periods of time – weeks or months, whereas close to a planet, where the gravitational field is stronger, a low-specific-impulse, high-thrust configuration – more akin to a chemical rocket – might do better.

VASIMR propulsion will likely find applications in deep-space interplanetary missions, as well as in cargo transfer to and from the moon. It has even been proposed for deflecting asteroids that may pose an impact threat to the Earth – by attaching a VASIMR drive to the asteroid to nudge it gradually onto a new trajectory.

Sailing on light

Another low-thrust, long-burn propulsion concept that's attracting interest is the solar sail. Like a sailing ship blown across the surface of the ocean by the wind, a solar sail spacecraft would deploy a giant sheet of reflective material, allowing it to be blown across the solar system by the light streaming out from the sun.

This relies on a concept from physics called 'light pressure'. In the 1860s, Scottish physicist James Clerk Maxwell published his unified theory of electricity and magnetism. It predicted that light – which is a form of electromagnetic radiation – can develop a pressure, just like a gas, that's able to exert a force on solid objects. Russian physicist Pyotr Lebedev, in 1899, provided the first experimental demonstration of light pressure, recording the tiny force it induced using a delicate measuring device called a torsion balance.

Paradoxically, light pressure is perhaps most easily visualized through one of the more inscrutable theories in modern science: quantum physics. This describes fundamental particles of matter and their interactions between

> ❘ **This is history in the making – LightSail 2 will fundamentally advance the technology of spaceflight.** ❜
>
> BILL NYE
> CEO, The Planetary Society (2019)

one another. It has some quite weird consequences, one of which is an effect known as wave–particle duality. This says that objects traditionally thought of as particles (such as protons and electrons) can also behave like waves and, conversely, that entities we normally think of as wavy (e.g. a beam of light) can act like a stream of solid particles. From this perspective, the sunlight shining on a solar sail is like a torrent of particles drumming down upon it, each imparting a tiny bump of momentum that pushes the sail forward.

The resulting acceleration is tiny, but quickly accumulates. A sail measuring a few hundred metres across could achieve speeds of 240,000 kph (150,000 mph) in about three years. Such a spacecraft would be able to reach Pluto from Earth in under five years. Compare that to NASA's New Horizons probe to Pluto – the fastest spacecraft ever flown from Earth – which, through a combination of chemical rockets and gravity assists, still took 9.5 years to reach the icy outer world.

Real solar sails were validated in a vacuum chamber on Earth between 2001 and 2005. The first flight came in 2010 when the Japanese space agency JAXA's IKAROS craft was able to demonstrate the viability of solar sails in interplanetary space, achieving a delta-v (the total change in speed imparted

THE SPACE ELEVATOR

Imagine a cable stretching from the surface of the Earth up into space that could be used to ferry payloads from the ground into orbit. This is the space elevator, a novel concept introduced by the visionary Russian space scientist Konstantin Tsiolkovsky at the end of the nineteenth century. It sounds a bit like the Indian rope trick. But the cable is, in fact, in a geostationary orbit around the planet's equator. It extends all the way down to the ground, where it's anchored, and it's capped off at the top with a large counterweight ensuring its centre of gravity circles the Earth at the geostationary altitude of 35,786 km (22,236 miles). In some sense, it's the centrifugal force of the circling counterweight that holds the entire structure up. Why haven't we built one? The problem is that the cable would be so heavy it would be unable to support its own weight. Even one made from carbon nanotubes, the strongest and lightest material we have, would break once its length exceeded 10,200 km (6,338 miles).

by a rocket engine, see Chapter 2) on their prototype sail of 100 m/s (330 ft) over a period of six months. Most recently, a SpaceX Falcon Heavy rocket carried LightSail 2 – a Kickstarter-funded solar sail built by the not-for-profit Planetary Society – into orbit around the Earth. Controllers

> **The space elevator will be built about fifty years after everyone stops laughing.**
>
> ARTHUR C. CLARKE
> (1984)

were able to raise and lower the altitude of the craft's orbit by manoeuvring the sail with respect to the sun's light.

This is how a solar sail could actually fly inwards towards the sun. Earth is on a solar orbit and so anything launched on a rocket from Earth finds itself on a similar path. With the sail edge-on to the sun, the spacecraft receives no thrust from sunlight and so maintains its orbit. But then tilting the sail forward allows sunlight to fall on its front face, slowing it down, and causing its orbit to spiral in gradually. Likewise, tilting the sail backwards, to illuminate its rear face, causes it to speed up, making its orbit wider and sending the spacecraft on a path towards the outer solar system. To maximize the thrust harnessed from the sun, solar sails are made from highly reflective materials, usually a thin plastic film such as mylar or polyimide, coated with a layer of aluminium.

Powered by light, the fastest thing in the universe, solar sails are a bit like having a rocket that spits out the quickest exhaust jet permitted by the laws of physics. This means that their gradual acceleration, acting over long periods of time, can ultimately propel them to terrifically high speeds, significant fractions of the speed of light. That's one of the reasons, as we'll discover later, why solar sails are one technology that could ultimately take us to the stars.

09 THERE'S NO PLACE LIKE SPACE

'You need to live in a dome initially, but over time you could terraform Mars to look like Earth and eventually walk around outside without anything on ... So it's a fixer-upper of a planet.'

ELON MUSK (2012)

The ISS has been circling the Earth above our heads since 1998, and it offers a glimpse of where the human presence in space may ultimately be heading. As we grow more adept at overcoming the challenges that space travel presents, so we can expect to see space stations constructed increasingly further afield. In Chapter 7, we looked at the Lunar Gateway – NASA's plan for a station in orbit around the moon, which could take shape as soon as the late 2020s. Plans are afoot for crewed bases on the surface of the moon and, perhaps in the 2030s and 2040s, on other planets such as Mars – where a permanent presence will allow exploration and scientific discovery on a scale that has hitherto not been possible.

On the ISS, crews are rotated roughly every six months, to minimize the physical and psychological toll that living in space takes on the body. But as space stations and crewed bases appear in ever-more far-flung corners of the solar system, visits from the home world will necessarily become less frequent. These remote installations will cease to be mere outposts, and instead become more akin to settlements and colonies – self-sufficient communities of pioneers who have struck out from Earth to forge a new home, perhaps even a new nation state or political territory, far away on the high frontier.

Plans already exist for crewed bases on the moon and Mars, including the use of chemical processing plants that can manufacture oxygen and rocket fuel from water deposits and atmospheric gases. This kind of 'applied chemistry' approach could be taken further – an idea known as 'terraforming' promises to let settlers engineer the environment of another world to make it more Earth-like. Others have envisaged humans living in the depths of interplanetary space, in giant rotating habitats known as O'Neill colonies.

Yet living in space presents a raft of new issues to overcome. Already, meeting the daily needs of astronauts during their relatively brief stints on the ISS presents its own set of problems, some of which we already have ingenious

> **We have to be able to recreate in space habitats which are as beautiful, as earth-like, as the loveliest parts of planet Earth, and we can do that.**
>
> GERARD K. O'NEILL
> *The High Frontier* (1976)

solutions for – and others that engineers and space scientists are still working on.

Sustenance in space

After breathable air – which, as we saw in Chapter 4, is made on the station by passing an electric current through water to separate it into hydrogen and oxygen – the next thing a space traveller will probably require is a drink. Water is extremely heavy – a cubic metre of it weighs a metric tonne (1.1 US tons, 2,200 pounds) – and piping an endless supply from the surface of the Earth up to the ISS is not practical. Just in terms of drinking water, a six-person crew each consuming the recommended two litres (0.5 gallons) per day would, between them, get through almost 4.5 tonnes (5 US tons) of water every year.

Instead, the station – and any other permanently occupied space habitat – uses a closed-loop purification system to recycle around 93 per cent of the water that the astronauts use. That includes waste washing water, moisture from the air – and, yes, urine. Water is recovered from urine through a distillation process, which separates the water from other chemicals that boil at different temperatures. It's also centrifuged to separate out gases. The

WHEN YOU'VE GOTTA GO …

Of all life's necessities, using a toilet without the luxury of gravity to 'help things along' is one of the more interesting aspects of living in space. Modern zero-gravity toilets use airflow from fans to direct the flow of waste products. Faeces are dried and compressed for later disposal. Urine – at least on the space station, where water is precious – goes into the purification system for recycling. Absence of gravity means that a rubber-gloved hand may be required to assist the elimination of solid waste from the body. Zero-g also removes much of the physical urge to 'go', meaning astronauts may need to schedule their visits in order to avoid accidents. And even then, the system isn't totally reliable – floating spheroids of urine, as well as bits of excrement, can and do escape into the crew areas. Even on the moon and Mars you won't get the full terrestrial experience. Mars is approximately one-third Earth gravity, while on the moon the natural downforce is just one-sixth of what you're used to on the pot back home.

water recovered in this way is combined with other waste water. This all goes through a processor to remove solid contaminants, and then passes through additional stages of filtration and chemical treatment to eliminate toxins and microorganisms. Finally, the purity is electrically tested, and any water not making the grade is thrown back in and processed again. It may sound gross, but in fact the drinking water on the ISS is purer than the stuff that comes out of most domestic taps.

> **Peeing or pooping in space is now a lengthy process, involving a fan, a targeting system, and a fair amount of prayer.**
>
> MARY ROBINETTE KOWAL
> (2019)

Culinary experiences on the ISS are a far cry from the tubes of pureed food eaten by Yuri Gagarin during his historic circuit of the Earth in 1961. Today, full meals are transported to the ISS in vacuum-sealed bags, and heated in the station's galley. Drinks are shipped in powdered form and then rehydrated. Fresh fruit and veg are an occasional morale-boosting treat, though must be eaten quickly before they spoil.

Crumbs are a no-no in microgravity – floating away and potentially contaminating instruments and equipment. So bread, cake and other crumbly foodstuffs are normally off the menu. The zero-gravity environment makes it difficult for fluids to drain from a person's head, leading to congestion, which interferes with smell and taste. For this reason, strongly flavoured foods tend to be popular with astronauts. This effect makes some foods taste and smell radically different in space. A plan to send some cream sherry up to the US Skylab astronauts in the 1970s had to be abandoned after it was discovered that the aroma of sherry in zero-g triggers the gag reflex.

Good health

Generally speaking, the drinking of alcohol in space is now prohibited for safety reasons. There have been some notable exceptions, though. During the Apollo 11 lunar landing in 1969, Buzz Aldrin took communion in the *Eagle* module, drinking wine. While in the 1990s, Russian cosmonauts smuggled vodka, cognac and other spirits onto the Mir space station, which were imbibed freely, much to the chagrin of visiting NASA astronauts, who remained abstinent. If you really must, spirits and non-fizzy wines are the medicines of choice in space. Beer, and other carbonated drinks, are not a good idea at all. In zero gravity, gas and liquid cannot separate out in your stomach as they

> **A year is a long time to live without the human contact of loved ones, fresh air, and gravity, to name a few.**
>
> SCOTT KELLY
> NASA astronaut (2016)

can on Earth, meaning that burping becomes a vomit-like experience known as a 'wet belch'.

Even if you lay off the beer, nausea in space remains a big issue, with over 60 per cent of space travellers experiencing motion sickness after leaving Earth. The good news is that space sickness normally only lasts for one to two days, and so shouldn't pose a long-term problem for those living in zero gravity.

Space colonies will ultimately need to be self-sufficient and produce their own food. Experiments such as the Vegetable Production System on the ISS have proven that it's possible to grow veg in microgravity. And if these crops can grow in space, then, given the right nutrients and atmospheric composition, they should grow on other planets too. At least to begin with, space colonists will need to adopt a vegan diet, as meat and other animal-based foods require too much in the way of resources to produce.

Research has shown that Mars, for example, already has many of the nutrients in its soil required to grow vegetables, and its atmosphere is 95 per cent carbon dioxide – the gas that plants require to generate energy via photosynthesis. The problem would be the extreme cold on the planet, meaning that crops there would need to be cultivated in heated greenhouse-like modules.

Even if you take care over nutrition, staying fit and well in space isn't straightforward. Human beings have evolved to thrive in Earth's gravity – if we want to live in space,

STAY COOL

Without the protection of an atmosphere, the temperature in near-Earth space varies wildly, from 120 °C (250 °F) in sunlight to −160 °C (−250 °F) in the shade. On the International Space Station, this demands some fairly heavy-duty thermal engineering to maintain a comfortable environment for the astronauts to live and work in.

On Earth, heat is wafted around by the atmosphere in processes known as conduction and convection. But in the empty vacuum of space it can only take one form: radiation. That means the layers of fluffy fibres you might insulate your loft with aren't much use here. Instead, reflective silver sheets are the best way to prevent heat from entering or leaving. That's why the ISS is clad with aluminized Mylar.

With all of its electrical equipment kicking out heat, keeping the ISS cool is the main challenge. Water is circulated on a loop around the station, passing through 'heat exchangers' – devices a bit like reverse central-heating radiators, that soak up heat from the crew areas and dump it into space.

or on other worlds with different gravity, then we need to establish the consequences for our health. In 2015, NASA conducted a unique and fascinating experiment. Astronaut Scott Kelly was sent into space for a marathon one-year stay on the ISS while his identical twin brother, Mark, remained on Earth. The brothers were both subjected to a battery of medical tests before and after the flight. Among the differences found after Scott's return to Earth were genetic alterations, possibly caused by radiation in space, and a small decline in cognitive ability. Most of these changes, though, eventually reverted with time, as Scott re-acclimatized to living on Earth.

Other studies have highlighted the impact of space on astronauts' mental health. The cost and difficulty of putting people into space naturally means that they are worked hard once there, and this, coupled with the limited options for recreation, plus the necessity to live and work in a confined space with the same people for weeks, months and sometimes years, can take its toll. In 1973, astronauts aboard the US Skylab space station reportedly went on strike after forty days of punishing schedules and little respite, refusing to communicate with mission control. Several Russian missions in the 1970s were plagued by interpersonal issues between crewmembers. The Soyuz-21 flight in 1976 was even cut short and a subsequent investigation concluded that the cosmonauts had experienced stress-related hallucinations.

NASA puts astronaut candidates through many hours of psychological screening, while those in flight aboard the ISS receive fortnightly psychiatric assessments by medical professionals on the ground. The station also keeps

> **All the conditions necessary for murder are met if you shut two men in a cabin measuring 18 feet by 20 and leave them together for two months.**
>
> VALERY RYUMIN
> cosmonaut, Salyut 6 (1980)

a supply of antidepressant, antipsychotic and anti-anxiety medications, and even has the facilities to physically restrain a crewmember if necessary.

While most of us might expect astronauts to have strong, resilient, outgoing personalities, it's been suggested that the ideal crewmember for long-haul missions to Mars and beyond might actually be more introverted. Those with a reduced need for social stimulation, who enjoy solitude, may turn out to be far more stable psychologically and better equipped to deal with the isolation of living billions of miles from home.

Islands in the sky

The greatest impact of long-term exposure to reduced gravity on the human body is borne by the musculoskeletal system. Without gravity to turn your body mass into weight, muscles soon weaken and atrophy, and bones become porous and brittle. In as little as two weeks, an astronaut can lose 20 per cent of their muscle mass. And that includes the heart, which can lead to circulation problems on return to Earth. Bone mass is lost at the rate of around 1.5 per cent per month. The solution to both problems is for astronauts in space to take regular exercise. And not just for five minutes – astronauts on the ISS do two hours of exercise per day on a treadmill, a weights machine and an exercise bike, with bungee cords simulating the effect of gravity.

One way that's been proposed to combat physical deterioration on crewed deep-space flights, and in space colonies, is to simulate gravity by making the spacecraft or space station rotate. Just like clothes stuck to the inside of a washing-machine drum by centrifugal forces, astronauts aboard such a rotating spacecraft would experience a force sticking them to the inside of the craft. And if the rotation speed were set just right, this force could be made equal to the gravitational pull at the Earth's surface.

In the 1970s, the American physicist Gerard O'Neill imagined vast colonies of humans living in space on the insides of rotating cylinders, made out of building stock mined from the moon, asteroids and other planets – and the rotation providing artificial gravity. Each cylinder was 8 km (5 miles) in diameter and some 30 km (20 miles) in length. The surface would consist of six equal-area stripes, stretching the length of the cylinder, three of which are land areas, with the other three, placed alternately between the land stripes, being giant windows to let in sunlight. By rotating about once every two minutes, O'Neill's cylinders could simulate Earth gravity on the inside.

The land areas would be literally that – not sci-fi space modules, but actual land, where inhabitants could grow trees and crops, and even farm livestock. This way, colonies could gain a degree of independence, requiring minimal shipments of supplies from Earth. The land areas would also allow inhabitants to walk 'outside', escaping much of the confinement and 'cabin fever' associated with more traditional space habitats, which can be so harmful to astronauts' mental wellbeing. In fact, O'Neill believed that his artificial habitats would be not simply the equal of Earth but, in his own words, 'far more comfortable, productive and attractive'. And he believed that they could be built with 1970s technology, let alone that available today.

In O'Neill's vision, the cylinders circle the sun at the L5 Lagrange point (see Chapter 3), at the same distance as the Earth from the sun but 60 degrees behind the planet in its orbit. The cylinders are connected together in pairs, counter-rotating so as to cancel out the angular momentum. The rotation axis of each cylinder pair points in towards the sun, with a large curved mirror behind each window strip to reflect light in. The inside is filled with an oxygen–nitrogen atmosphere, at roughly half of the Earth's atmospheric pressure at sea level. O'Neill calculated that this would be sufficient to block harmful cosmic radiation and also permit the cylinders to sustain their own weather. The windows would be made of very many individual glass panels, so that if any were punctured by meteoroid or space-debris impacts they could be replaced without risk of the cylinder depressurizing.

> **If you don't like airline food you'll probably have the same impression of space station food. I would not fly to space for the food.**
>
> CHRIS HADFIELD
> astronaut (2014)

At a news conference in 2019, Jeff Bezos, the founder of private space launch company Blue Origin (and Amazon. com) announced his support for the concept of O'Neill cylinders rather than settlements on other planets for the ultimate human colonization of space. And he has a point. The gravity and weather inside the cylinders can all be controlled, rather than forcing colonists to grapple with an alien Mother Nature. And there'll be no other geological curve balls to deal with – like earthquakes or volcanism.

Terra forma

In the far future, there could be another even more radical possibility waiting in the wings – one that makes planetary settlement more appealing. Called terraforming, it involves transforming the environment of a barren world on a global scale to make it like Earth. Literally, a home away from home. Colonists on a terraformed world have no need for spacesuits, or for sealed habitats in which to live and grow food – they would engineer their own planetary biosphere.

The concept, which literally means 'Earth shaping', was put forward by Carl Sagan in an article he published in the journal *Science* in 1961 – having been explored by science-fiction writers even earlier. Sagan's original idea was to add algae to the atmosphere of Venus to try and seed the development of organic chemistry on the planet. Others have suggested launching a space parasol to sit in orbit between Venus and the sun and shade the planet from the solar heat, allowing its atmosphere to cool. However, its proximity to the sun, and the fact that the planet is in the grip of a runaway greenhouse effect, leading to extreme temperatures and crushing atmospheric pressure at its surface, make terraforming Venus seem like a truly mammoth undertaking.

A more suitable target might be Mars. Evidence suggests it's very likely that, billions of years ago, Mars had a warm, wet climate, able to support life. Of all the worlds in the solar system, Mars is probably the most Earth-like today. The principal differences are its atmosphere (which is very thin and made mostly of carbon dioxide, compared to the oxygen–nitrogen atmosphere on Earth), its surface temperature (which averages around –60 °C, or –80 °F) and its lack of an appreciable magnetic field. The Earth's magnetic field acts as a shield against cosmic radiation, and in particular serves to deflect the 'solar wind', the stream of charged particles constantly billowing out from the sun. It's likely that the solar wind was a factor in eroding and stripping away the atmosphere on Mars.

Any plan to terraform the planet would need to remove these differences. One way is to trigger a greenhouse effect on Mars – much like the greenhouse effect here on Earth, where the atmosphere is trapping more heat from the sun than the planet can radiate away. Except that, whereas on Earth the greenhouse effect is a catastrophe in waiting, making the planet dangerously warm and risking a climate runaway of the sort seen on Venus, on Mars it would be a good thing, thawing the planet and making it habitable. Once Mars began to warm, the frozen reserves of water and CO_2 at the planet's poles would melt, enhancing the greenhouse effect and heating the atmosphere further.

One way to start this process might be to transport powerful greenhouse gases, such as sulphur hexafluoride (chemical formula SF_6, which is nearly 24,000 times more potent as a greenhouse gas than CO_2) to Mars and dumping them into the planet's atmosphere. This would still require hundreds of thousands of tonnes of SF_6, which would need to be manufactured and hauled all the way to the Red Planet. Another chemical that's been suggested is ammonia. This is also a strong greenhouse gas, but is plentiful in the icy minor planets that circle the outer solar system. One possibility might be to alter the orbit of one of these bodies, perhaps by attaching a super-efficient electric rocket engine, to place the body onto a collision course with Mars.

But the effort's all wasted if the planet doesn't have a magnetic field to help hold on to its new atmosphere. It's been proposed that a network of superconducting electromagnets could generate an artificial field around Mars large enough to protect its atmosphere. Electromagnets are coils of wire that, because electricity and magnetism are just different aspects of the same thing, generate a magnetic field through their core when a current is passed through the wire. Making the wire from a superconducting material greatly increases the size of the current permitted, in turn boosting the strength of the field.

Another option still was proposed in 2017 by Jim Green, director of NASA's Planetary Science Division.

> **Once we become a multiplanet species, our chances to live long and prosper will take a huge leap skyward.**
>
> DAVID GRINSPOON (2004)

He suggested placing a similar magnetic generator at the L1 Lagrange point (see Chapter 3) in space between Mars and the sun – from where it could deflect the flow of the solar wind around the planet. Green envisages the shield

IRONMAN IN SPACE

The record for the longest continuous stay in space is held by Russian cosmonaut Valeri Polyakov who, between 8 January 1994 and 22 March 1995, spent a total of 437 days aboard the Earth-orbiting Mir space station. Polyakov is a physician specializing in space medicine. He was selected to become a cosmonaut in 1972, though didn't make his first spaceflight until 1988. He volunteered for the marathon stay on Mir to assess the impact of long-duration spaceflight on the human body. The biggest impact was found to be on his mental wellbeing and mood. Polyakov's physical fitness fared remarkably well, with no long-lasting side effects. He even made a point of walking from his capsule after landing (most returning cosmonauts are carried) – pretty good after fifteen months in zero gravity. Polyakov did also hold the record for the longest accumulated time in space. That honour now goes to his fellow cosmonaut Gennady Padalka, who has racked up a total of 879 days over six missions.

as a small inflatable structure. Computer simulations suggest that the plan could work, guarding the planet from the hail of particles in the solar wind and allowing it to retain enough of an atmosphere to increase its average temperature by around 4 °C (7 °F) – sufficient to begin the melting of the planet's polar cap.

Terraforming, and the colonization of other worlds in general, has thrown up ethical questions. There are inevitable parallels to be drawn with the first European settlers to arrive in Australia or the Americas. Let's hope, though, that by the time terraforming becomes technologically feasible, we will have learnt our lesson to respect indigenous life – and only exploit, colonize or terraform worlds, and call them our own, once we have established beyond doubt that they are lifeless and barren, and available for human habitation.

10 HOW TO REACH THE STARS

'Across the sea of space, the stars are other suns.'

CARL SAGAN, *COSMOS* (1980)

In 2012, Voyager 1 became the first spacecraft from Earth to enter interstellar space. Launched in 1977, it flew past Jupiter and Saturn before being slingshotted off onto a trajectory taking it out of the solar system. In August 2012, at a distance of 18 billion kilometres (11 billion miles) from the sun, mission controllers reported that the probe had officially crossed the heliopause. This is the boundary formed where the solar wind from the sun collides with the nebulous gas making up the interstellar medium – it marks the edge of the solar system and the beginning of interstellar space.

In November 2019, its sister craft Voyager 2 also crossed into the domain of the stars. The Voyagers are part of an elite group of five spacecraft that have all achieved the necessary speed to escape from the sun's gravitational field – the others being Pioneer 10 and 11, and the New Horizons probe, which flew past Pluto in 2015.

When Voyager 1 left the solar system, it was 121 times further from the sun than the Earth is. But this is a hop and a skip compared with the distances to the other stars in our Milky Way galaxy. The nearest star to the sun, Proxima Centauri, is still 4.3 light-years away. That's about 4 trillion km (2.5 trillion miles), or more than 270,000 times the Earth–sun distance. It is a truly immense void to cross. Moving at 17 km/s (11 miles per second), Voyager 1 isn't exactly hanging around, but even at that speed – and even if it were heading in the right direction (it's not) – it would take the craft 70,000 years to reach Proxima Centauri.

This raises natural doubts as to whether robotic space probes, let alone human astronauts, will ever be able to make the journey and explore other star systems. And yet there are some daring – some might say fanciful – concepts on the drawing board that could, one day, make interstellar space travel a reality.

Ad astra

In 2016, Russian entrepreneur Yuri Milner and the late cosmologist Professor Stephen Hawking announced Breakthrough Starshot, an ambitious project to send a flotilla of tiny robotic space probes to Proxima Centauri.

The probes would visit Proxima Centauri b, a planet orbiting the star that was discovered in 2016 by an international team of astronomers using the telescopes of the European Southern Observatory in Chile. Proxima b is interesting because it's thought to be a rocky, Earth-like world, and it's within the habitable zone of its star – the region where conditions are suitable for life.

The spacecraft would be powered using solar sails to hitch a ride on light (see Chapter 8). Rather than using light from the sun, though, it's proposed that the craft would be propelled by an Earth-based array of high-power laser beams. This idea was first put forward in the 1970s by the American physicist and science-fiction writer Robert Forward, as a way to operate a solar sail in the darkness of interstellar space. Breakthrough Starshot calls for a laser array with a total output power of 100 gigawatts. That's a lot of power. To put it into perspective, the world's largest nuclear power plant, the Kashiwazaki-Kariwa plant in Japan, generates almost 8 gigawatts.

The spacecraft themselves are tiny, each measuring a few centimetres across and weighing a gram (0.035 ounces) or less. Called StarChip, a number of prototypes were successfully

> Nothing travels faster than the speed of light with the possible exception of bad news, which obeys its own special laws.
>
> DOUGLAS ADAMS
> *Mostly Harmless* (1992)

flown in low Earth orbit in 2017. Each craft packs four tiny cameras, a computer, miniature thrusters, a plutonium battery, a radio transmitter and a 16-square-metre (170-square-foot) reflective lightsail.

The current plan calls for a mothership to be launched into a high-altitude Earth orbit by a rocket. From here, it would deploy a swarm of the StarChip spacecraft. The laser array then illuminates each craft for approximately ten minutes. The thrust delivered is small – enough to lift just a few hundred grams (a few tens of ounces) from the Earth's surface – but in space, pushing against such a lightweight payload, that force can accelerate the craft ridiculously quickly, propelling them to between 15 per cent and 20 per cent of the speed of light. At this speed, the journey to Proxima b would take twenty to thirty years, plus another 4.3 years for its findings – travelling as a radio transmission at the speed of light – to be beamed back to Earth.

Travelling so fast, collisions with even so much as tiny particles of dust could be catastrophic to any one of the craft. And so, to build some redundancy into the mission plan, it's envisaged that around around 1,000 StarChips would be launched from the mothership.

Critics have suggested that many technologies will have to increase tenfold and then some for Breakthrough Starshot to be successful. However, the team of scientists advising the project – including astrophysicist Avi Loeb, Nobel prize-winner Saul Perlmutter, and Britain's Astronomer Royal Martin Rees – believe it can be done.

Self-replicating probes

Breakthough Starshot is perhaps the most promising project there's been to send a spacecraft from our star system to another. But then what? When the nearest star takes over twenty years to reach, working through the rest of our galaxy's 100 billion stellar targets one at a time like this is going to take a while. One solution that's been put forward is to develop a robotic probe capable of flying to other stars and gathering data about them and their planets autonomously, with no direct control from Earth.

> **Earth is a wonderful place, but it might not last for ever. Sooner or later, we must look to the stars. Breakthrough Starshot is a very exciting first step on that journey.**
>
> STEPHEN HAWKING (2016)

Rather like a primitive sort of lifeform, the probes would also be capable of self-replication. By commandeering resources found on asteroids and other worlds, utilizing them through mechanical, electronic and even nano-engineering – rearranging materials at an atomic or molecular level – the spacecraft and their descendants would spread across the galaxy, gathering data and knowledge and all the while creating ever more copies of themselves as they went.

These spacecraft, or rather the idea of them – no one's ever actually built or launched one – are known as von Neumann probes, after the Hungarian-American polymath John von Neumann. In the 1940s, von Neumann proved mathematically that self-reproducing machines are the most efficient way to explore space. The idea was developed in 1980 by the American nanotechnologist Robert A. Freitas, who carried out detailed calculations demonstrating the feasibility of von Neumann probes as a method for exploring the galaxy.

A study in 2013 suggested that the probes could employ a stellar analogue of the gravity-assist manoeuvre used to catapult spacecraft across the solar system. By swinging around fast-moving stars they could boost their speed dramatically, making it possible for a swarm of the probes to explore the entire Milky Way galaxy in around 10 million years. That may seem like a long time, but it really is the blink of an eye in astronomical terms.

Some scientists have cited the apparent lack of any alien von Neumann probes plying our solar system as proof that there is no other intelligent life in the galaxy. And yet we've seen that even our own technology today is capable of producing robotic spacecraft so small that they could evade detection. The Breakthrough Starshot spacecraft, for example, are tiny – just centimetres across. It's quite possible they could be here observing us and we'd never know.

> If we're going to have any chance of sending stuff to other star systems, we need to be laser-focused on becoming a multi-planet civilization. That's the next step.
>
> ELON MUSK (2014)

Getting robots across the galaxy is one thing, but how might humans make the journey? Science fiction often depicts astronauts being placed into hibernation for the duration of these long-haul spaceflights. In 2019, ESA conducted a study into using chemically induced torpor to slow the human body's metabolic rate by as much as 75 per cent, greatly reducing the demands on life support. This theoretical study found that hibernating the crew on a flight to Mars would enable mission planners to reduce the mass of the spacecraft by as much as a third, because of the reduced need for water, food and oxygen. And this would make propelling the vessel on its journey from Earth to the Red Planet a lot easier. At present, however, the technology is only being considered for interplanetary missions. If this proves successful, then flights beyond the solar system will be the logical next step.

Breaking the law

In science fiction, human travellers make journeys across the galaxy often by casually exceeding the speed of light. Although it might seem fantastical, travelling at faster-than-light speed may not be quite as impossible as Einstein led us to believe. There are a couple of ways it could be done, both of them indirectly down to Einstein and his theories. Although, as we'll see, envisaging either of them as an actual reality any time soon may require some imagination on the part of the reader.

> **By a purely local expansion of spacetime behind the spaceship and an opposite contraction in front of it, motion faster than the speed of light ... is possible.**
>
> MIGUEL ALCUBIERRE
> physicist (1994)

The main obstacle to travelling faster than light is Einstein's special theory of relativity, of 1905. This is a theory governing the motion of moving bodies. For the most part, its predictions agree with those of Newton's earlier laws of motion, yet at very high speeds Newton and Einstein go their separate ways. And it's Einstein's theory that's borne out by experiment.

Special relativity was inspired by James Clerk Maxwell's theory of electromagnetism. Among the consequences of Maxwell's theory, formulated in the 1860s, is the assertion that the speed of light is a fundamental constant of nature, and is therefore the same whatever frame of reference it's measured in. Einstein reasoned that this was at odds with the normal concept of relative motion. For example, if you're in a car travelling at 50 km/h (31 mph) following another car travelling at 30 km/h (19 mph), then, relatively speaking, you're travelling towards the other car at 20 km/h (12 mph). Maxwell's theory was suggesting that this didn't apply to light – that however fast you travelled, light would always appear to move at the same speed.

The special theory of relativity was built with this as one of its cornerstone assumptions. And from it came some dramatic consequences – such as the idea that, in a moving frame of reference, lengths contract and clocks tick more slowly, and that energy and mass can be considered equivalent, linked by the famous formula $E=mc^2$. All of these were later validated by experiment. But there was another prediction too – that as a body accelerates, its mass increases, slowly at first but then more rapidly, until, at the speed of light, the mass becomes infinite. Making an infinite mass go any faster requires an infinite amount of energy, from which Einstein inferred that nothing could travel faster than a beam of light. Special relativity enforces light speed as the maximum speed limit of the universe.

That's fine within the special theory, and the special theory alone. But in 1915, Einstein published his general theory of relativity. This incorporated gravity into the model, by bending and stretching the flat space and time of special relativity. And this opened the door to some very strange possibilities indeed.

Hyperspace

In 1916, the year after general relativity was published, an Austrian scientist called Ludwig Flamm used the theory to build the first mathematical model of what scientists today call a wormhole – a tunnel through space and time. Flamm found that if you take the mathematical solution to general relativity describing the space around a central gravitating body – like a star, or a black hole – then the space can be extended into the core of the body and beyond, through to a brand-new region of space.

The term 'wormhole' to describe these theoretical objects was coined in the 1950s by the American physicist John Wheeler, who likened them to the tunnels made by a worm burrowing into an apple – the straight distance from one side to the other being shorter than the distance around the apple's circumference. And this is where wormholes could come into their own for space travel. For while they cannot physically accelerate a spacecraft to faster-than-light speeds, they could open up cosmic shortcuts, linking far-flung destinations by a network of back alleys through higher-dimensional hyperspace, to dramatically reduce the travel times between distant star systems.

The problem is holding the wormhole open. The mathematics predicts that a wormhole's throat behaves like a rubber tube, and the effect of gravity is to then try and pinch it shut. General relativity might just hold the solution though. As a theory of gravity, it determines the gravitational force – or equivalently the curvature of space – generated by any kind of substance that you care to imagine.

Physicists have come up with a theoretical kind of material that they call 'exotic matter', which actually has negative pressure – meaning that if you were to pump it into a balloon or the tyres on your car, they would get flatter. The key thing about exotic matter, though, is that it generates 'antigravity' – so that if you thread a wormhole with enough of the stuff it can generate a repulsive gravitational force sufficiently strong to hold the throat open while a spacecraft travels through.

Sadly, exotic matter isn't all that easy to come by. But small quantities have been produced in the lab, in a phenomenon called the Casimir effect. This was discovered in 1948 by a Dutch scientist called Hendrik Casimir.

He found that two metal plates separated by just a few millionths of a metre in a vacuum are pulled together slightly by the negative pressure of exotic matter between them.

The exotic matter is created by what physicists call 'vacuum fluctuations' – scores of tiny subatomic particles popping in and out of existence over very short times. Quantum theory says that these particles can equally well be thought of as waves. In the same way that the note played on a guitar is partially determined by the length of the strings, only waves of a certain wavelength can fit between the plates. Back in particle terms, this means there are fewer particles knocking around between the plates than outside of them. This makes the pressure between the plates lower than it is outside. But if the space outside the plates is in a zero-pressure vacuum, then the space inside must have less than zero, or negative pressure.

Engage!

The other science-fiction space-travel staple that has been examined from the standpoint of science fact is warp drive – the notion that a spacecraft could move faster than light by deforming space and time.

If special relativity ruled supreme, spacecraft would be constrained to travel at less than the speed of light, 300,000 km/s (186,000 miles per second). In 1994, physicist Dr Miguel Alcubierre, of the University of Wales College of Cardiff, calculated how a warp drive might actually work within the context of general relativity. Instead of trying to move the spacecraft through space, his approach was to bend and stretch the space around it to form a wave that sweeps the craft along to its destination.

Like a wormhole, this propulsion system also requires anti-gravitating exotic matter. Alcubierre found that if the exotic matter were to be arranged around the spacecraft in just the right way, its gravitational effect would cause the space in front of the craft to shrink rapidly, while the space behind it would expand at exactly the same rate, carrying the piece of space in between, containing the craft, along to its destination at a speed faster than light.

> **You can't go faster than the speed of light, but what you can imagine doing is effectively twisting space-time so that it looks like you're moving faster than the speed of light.**
>
> SEAN CARROLL
> physicist (2018)

Once again though, the stumbling block with a warp drive is turning the science – which is generally solid – into

ENGAGE ANTIMATTER DRIVE!

A staple of *Star Trek* and other science-fiction yarns, antimatter is appealing for spacecraft propulsion because of the large amount of energy it packs into a small mass of material. A particle of antimatter is like a particle of ordinary matter, but its electrical charge as well as other key properties are all reversed. When matter and antimatter meet, they annihilate each other, the mass of both particles being converted entirely into energy – a thousand times more than is released by the same mass of nuclear fuel. Antimatter-powered spacecraft would resemble the designs for nuclear rockets, described in Chapter 8, with energy from the annihilation either directly heating a propellant or being harnessed to generate a current that could drive an electric rocket, such as an ion or VASIMR engine. The antimatter is contained within the spacecraft using electric and magnetic fields. The principal drawback is that while nuclear fuels are naturally occurring on Earth, antimatter has to be manufactured. And the process is costly, inefficient and, at present, only able to produce a few atoms of the stuff at a time.

real-world engineering. Both wormholes and Alcubierre's warp drive design demand more exotic matter than we can feasibly produce – typically around a planet-sized mass of the stuff. By comparison, the quantity generated in the Casimir effect is minuscule.

It's interesting that Casimir's experiment demonstrates exotic matter can be created artificially within the laws of physics – all that's missing is the not-inconsiderable technology required to mass-produce it. Yet this capability seems a long way beyond our grasp.

In 2008, the US Department of Defense investigated warp drive, wormholes and other speculative ideas from science. When the report was released in 2018, Caltech physicist Sean Carroll commented: 'There is something called a warp drive, there are extra dimensions, there is a Casimir effect, and there's dark energy – all of these things are true. But there's zero chance that anyone within our lifetimes or the next 1,000 years is going to build anything that makes use of any of these ideas.'

> I think that the future of the human race is to spread through the universe, and now is the time that we should be laying the foundations for that.
>
> KIP THORNE (2015)

That said, perhaps we should take a slightly more philosophical view. Ideas such as heavier-than-air flight and spaceflight itself were both dismissed as preposterous by the great scientists of their time – in some cases just a few years before the concepts were successfully demonstrated. As the great scientist, science-fiction author and visionary Isaac Asimov sagely advised: 'Your assumptions are your windows on the world. Scrub them off every once in a while, or the light won't come in.'

GLOSSARY

Ablation A physical process where material is removed from an object by gradual erosion. Ablative heat shields on spacecraft carry away heat as the shield chars and pieces break off.

Acceleration Any change in an object's state of rest or motion at constant speed. Acceleration is caused by the action of a force.

Aerobraking A manoeuvre whereby a spacecraft reduces its speed by skimming through the upper layers of a planet's atmosphere.

Airbag One system for landing robotic spacecraft on the surface of other planets involves surrounding them with a cluster of inflatable airbags to cushion their impact with the ground.

Angular momentum A rotational equivalent of momentum, given by the angular speed of an object and the distribution of the object's mass around its centre of gravity.

Aphelion The point in the orbit of a planet or spacecraft around the sun at which its distance from the sun is greatest.

Breakthrough Starshot A project to use solar sails to send miniature space probes to another star system.

Delta-v The total change in velocity imparted to a spacecraft by the firing of a rocket engine.

Drag force Resistance to motion experienced by any object moving through a fluid, such as an aircraft or spacecraft moving through the Earth's atmosphere.

Electric rocket A rocket engine that makes a high-speed exhaust jet by accelerating a charged gas using electric and magnetic fields.

Electromagnetism The unified theory of electric and magnetic fields developed by Scottish physicist James Clerk Maxwell during the 1860s.

Ellipse Stretch a circle in one dimension and the result is an oval shape known as an ellipse. Orbiting bodies, such as spacecraft, typically move on elliptical paths.

Escape velocity The speed at which a body must be blasted from the surface of a celestial object in order to escape that object's gravity. The escape velocity of the Earth is 11.19 km/s (6.95 miles per second).

Force Any kind of interaction with a physical object that causes the object to accelerate. Collisions, gravitational fields and rocket engines all induce acceleration by exerting a force on objects.

Freefall A state of motion in which an object moves in a gravitational field without experiencing the action of any other force. Examples include apples falling from trees and spacecraft in orbit.

G-force Acceleration expressed in terms of the acceleration due to gravity experienced by objects at the surface of the Earth. So, 5 g, for example, is five times Earth gravity.

General relativity Albert Einstein's theory of gravity in which the gravitational force manifests itself as curvature of space and time.

Geostationary orbit A satellite in a geostationary orbit circles the Earth at the same rate at which the planet spins, appearing to hang in the sky from the planet's surface.

Gravity assist A manoeuvre whereby an interplanetary spacecraft uses the gravity of another planet to pick up speed.

Ion engine An electric rocket where a gas of charged ions is accelerated with an electric field.

Isotope Different istopes of a chemical element have different numbers of neutron particles in their atomic nuclei, making them behave differently during nuclear reactions.

Kármán line The line delineating the boundary between Earth and space, located 100 km (60 miles) above the planet's surface.

Kepler's laws Three laws governing the motion of planets around the sun, deduced by German astronomer Johannes Kepler in the early seventeenth century.

Lagrange points Five points in the joint gravitational field of a two-body system where a third body can be placed and remain stationary relative to the other two.

Laser Short for 'light amplification by the stimulated emission of radiation', a laser is a device that generates a very narrow beam of intense light.

Launch escape system A rocket-propulsion unit to eject the crew capsule from a spacecraft in the event of a catastrophic emergency.

Launch window The interval permitted by the laws of physics for a space mission to launch to its destination.

Lightsail A variation on a solar sail but using directed laser light, instead of the light from the sun, to propel a spacecraft.

Low Earth orbit An orbit around the Earth with an altitude between 160 km (100 miles) and 2,000 km (1,240 miles).

Mach number The speed of a vehicle measured in multiples of the speed of sound. For example, Mach 2 = twice sound speed.

Micrometeoroid A tiny particle of rock in space, often moving so fast that it is capable of causing significant damage to spacecraft.

Momentum A property of a moving body defined as its mass multiplied by its velocity. The greater the momentum of an object, the more force it exerts in a collision.

Multi-staging A technique for reaching space where multiple rockets are stacked up on top of each other, boosting the total delta-v imparted to the payload.

NERVA Nuclear Engine for Rocket Vehicle Application (NERVA) was an early nuclear thermal rocket engine developed by NASA.

Newton's laws Three laws formulated by Sir Isaac Newton in the late seventeenth century, governing the behaviour of moving bodies.

Nuclear thermal rocket A type of rocket engine where propellant is heated by a nuclear reactor, rather than chemical combustion.

O'Neill colony A concept for a permanently occupied space habitat consisting of 30-km-long (20-mile-long) cylinders that rotate to generate artificial gravity.

Orbit A trajectory in the gravitational field of a massive object, such as a moon, planet or star.

Perihelion The point in the orbit of a planet or spacecraft around the sun at which its distance from the sun is smallest.

Radiation Any emission of high-energy subatomic particles or electromagnetic waves given off in a physical process.

Most of the harmful radiation in space consists of high-energy particles.

Radioisotope Thermoelectric Generator Often abbreviated to RTG, this is a device that uses the heat produced by the decay of a radioactive material to generate electricity. RTGs are used on space missions where solar power would be insufficient.

Sample-return mission Any space mission that returns samples of a celestial object to Earth for analysis in a laboratory.

Solar panel A device for converting light from the sun into electrical power – used by long-duration space missions, both crewed and uncrewed, to generate electricity.

Solar sail A spacecraft propelled by light (either from the sun or a laser) falling on a large reflective sheet resembling a sail.

Space debris Dead satellites, discarded rocket stages and pyrotechnic fragments all contribute to the growing cloud of hazardous debris in Earth orbit.

Space elevator An idea for reaching space using a cable stretching from the Earth's surface to a platform in geostationary orbit.

Space Shuttle A reusable space plane that was operated by NASA between 1981 and 2011.

Space sickness Nausea, headaches and disorientation experienced by many space travellers as they adapt to zero gravity.

Space station A permanently inhabited space platform, usually in orbit around a planet or moon.

Special relativity Albert Einstein's theory governing the dynamics of moving bodies. It extends Newton's laws to describe objects moving at close to light speed.

Specific impulse A measure relating to the amount of delta-v that can be wrung out of a rocket engine per unit mass of fuel burned.

Sub-orbital flight A flight on a parabolic arc up into space and straight back down again. The spacecraft does not gather enough speed to reach orbit.

Supersonic A term used to describe any object moving faster than the speed of sound.

Terraforming A future concept for engineering the climates of other planets to make them Earth-like and suitable for human habitation.

Thermal protection system A set of measures for shielding a spacecraft from the extreme heat generated on entering a planet's atmosphere from orbit or interplanetary flight.

Transfer orbit An orbital trajectory in the sun's gravitational field that charts a course from one planet in the solar system to another.

V2 missile The first long-range ballistic missile, and the first human-made rocket to reach space.

VASIMR A kind of electric rocket engine that uses techniques from nuclear fusion research to create a high-temperature plasma and focus it into an exhaust jet magnetically.

Von Neumann probe A concept for a fleet of self-replicating robotic spacecraft that could hop from star to star, gradually exploring the galaxy.

Warp drive A theoretical spacecraft-propulsion system that works by stretching and squeezing the fabric of space and time.

Wormhole A hypothetical tunnel through space and time that, it's been proposed, could serve as a shortcut, bridging the great distances separating the stars.

FURTHER READING

For a vivid account of the Apollo programme that took the first humans to the moon:

A Man on the Moon: The Voyages of the Apollo Astronauts, by Andrew Chaikin, Penguin (1994).

For a perspective on the US crewed space programme from Apollo 13 flight director Gene Kranz:

Failure Is Not an Option: Mission Control from Mercury to Apollo 13 and Beyond, by Gene Kranz, Simon & Schuster (2000).

For the story of how German genius Wernher von Braun drove the development of rocket technology in the USA:

Dr. Space: The Life of Wernher von Braun, by Bob Ward, Naval Institute Press (2005).

For a memoir by the Apollo 11 mission's Command Module pilot:

Carrying the Fire: An Astronaut's Journeys, by Michael Collins, Pan (2019).

For a guide to the Soviet space programme by an award-winning space historian and engineer:

Korolev: How One Man Masterminded the Soviet Drive to Beat America to the Moon, by James Harford, John Wiley & Sons (1999).

For a Q&A on what it's like to go into space, from someone who's been there:

Ask an Astronaut: My Guide to Life in Space, by Tim Peake, Arrow (2018).

For a comprehensive history of rockets and spaceflight, up to the beginning of the twenty-first century:

Spaceflight: The Complete Story from Sputnik to Curiosity – and Beyond, by Giles Sparrow, DK (2019).

For an under-the-bonnet look at how rockets work:

Space Rockets Owners' Workshop Manual: Space Rockets and Launch Vehicles from 1942 Onwards, by David Baker, J. H. Haynes & Co. (2015).

For a guide to the solar system as discovered (mostly) by robotic spacecraft:

The Planets, by Brian Cox and Andrew Cohen, William Collins (2019).

For a guide to how robotic space probes work:

Space Invaders: How Robotic Spacecraft Explore the Solar System, by Michel van Pelt, Copernicus (2007).

For a tour of some of the most fantastic spacecraft designs imagined in art and literature:

Spaceships: An Illustrated History of the Real and the Imagined, by Ron Miller, Smithsonian Books (2016).

For a slightly more technical read into the details of rocket propulsion and spaceflight:

It's Only Rocket Science: An Introduction in Plain English, by Lucy Rogers, Springer (2008).

For the low-down on settlements in space, from the man who invented them:

The High Frontier: Human Colonies in Space, by Gerard K. O'Neill, Space Studies Institute (1976).

For a guide to the future of the human species beyond planet Earth:

The Future of Humanity: Terraforming Mars, Interstellar Travel, Immortality, and Our Destiny Beyond Earth, by Michio Kaku, Doubleday Books (2018).

For the definitive guide to wormholes and relativity, from a distinguished physicist:

Black Holes, Wormholes and Time Machines, by Jim Al-Khalili, Routledge (2012).

INDEX